AF443646

Fuzzy Logic Applications in Engineering Science

Fuzzy Logic Applications in Engineering Science

by

J. Harris
*formerly of the National University of Science and Technology,
Zimbabwe*

A C.I.P. Catalogue record for this book is available from the Library of Congress.

ISBN-10 1-4020-4077-6 (HB)
ISBN-13 978-1-4020-4077-1 (HB)
ISBN-10 1-4020-4078-4 (e-book)
ISBN-13 978-1-4020-4078-8 (e-book)

Published by Springer,
P.O. Box 17, 3300 AA Dordrecht, The Netherlands.

www.springeronline.com

Printed on acid-free paper

To Marian
Also to Jennifer, Andrea, James and Julia

Table of Contents

Chapter 3. Materials Selection   31

Chapter 4. Hydrodynamic Lubrication   47

Chapter 5. Elasto-Hydrodynamic Lubrication   63

Chapter 6. Fatigue and Creep   73

Foreword

John Harris 1929–2003

In an engineering career which spanned almost fifty years, John Harris lost none of his enthusiasm for new fields of exploration. In his late sixties he began his research into fuzzy logic with his drive and sense of endeavour undimmed. At the time he was inaugural Chair of Industrial Engineering at the new National University of Science and Technology, Bulawayo, Zimbabwe, having dedicated almost his entire professional life to the development of educational opportunity in engineering science in sub Saharan Africa, a career which led him to Chair of the department of Mechanical Engineering at the University of Nairobi, Kenya, then subsequently to set up and chair the department of Mechanical Engineering in the University of Zimbabwe, Harare, thence to the Chair of the Mechanical Engineering department at the University of Malawi, returning for his last post to Zimbabwe, to the new university in Bulawayo.

The preface to this text, probably the last words my father ever wrote, reflects his belief in the relevance of fuzzy logic and its potential for applications at a functional level in society. He presents fuzzy logic as a theoretical perspective, a prism, through which he invites the viewer to share his understanding of industrial systems in our society.

The book was intended to be fourteen chapters. My father completed and edited the first nine chapters of this volume while fighting a titanic battle against disease. He chose to leave the expert care and comfort of St Michael's Hospice to return to die at home, where, against all odds and with the unfailingly patient and practical assistance of his son, James Harris, nine chapters were finalised. Chapters ten to fourteen inclusive existed in unedited form and some of this latter group of chapters include papers published previously in professional journals, which were intended to be reworked for inclusion. I am grateful to the Institution of Mechanical Engineers and to the Institution of Chemical Engineers for their permission to reproduce papers at chapters eleven, twelve and thirteen. The last five chapters therefore represent work in progress. When I came to work on the manuscript for chapter fourteen, the last of this volume, I found the pages suddenly empty of my father's pencilled edits.

Primarily, of course, this academic publication was created as an educational springboard text, to encourage others to take research forward and continue exploration of the fascinating interface between fuzzy logic and engineering.

I should like to thank all those who have helped and supported the efforts to bring this book to publication, including Derek Duffet for his kind assistance, Nathalie Jacobs and all at Springer Science and Business Media and my family, especially my mother, Marian Harris, to whom this book is principally dedicated.

Julia Harris
March 2005

Preface

Over the ages, there are many who have contemplated the perfect disc of the full moon in the night sky, the perfection of which we know to be an optical illusion made possible by the distance between the object and the observer. This is a factor which dissolves the imperfections that are apparent to the close observer. The perfect disc is an abstraction of the type which is of the essential nature of scientific and therefore engineering thinking. The abstractions enable precise relationships to be formulated without the hindrance of the complexity of imperfect detail. The human mind has a propensity to search for abstractions and also to classify and generalise them. The advance of civilisations, and more specifically here, science and engineering as we know it, would have been impossible without this phenomenon. It can, however, have penalties as in stereotyping, which have all too often in history lead to injustices.

Associated with the perfect abstractions of the mind is the concept of precision; that of flawless conformance to the ideal. This too has its roots in the philosophy of the Ancient Greeks and has a pervasive influence in many cultures, but especially in the western world where it has been a salient feature of scientific development. Another vital element of Ancient Greek philosophy which has governed thought not only in science but in many other fields such as law, medicine and theology, is that of classical; (Aristotelian) logic. This represents the strict classification of real or abstract objects into sets of which they are either perfect members or not.

Judging by the history of the past few centuries, science and engineering have been remarkably successful in the application of these principles. The fruits of success have been the results of the influence of the concepts of abstraction, precision and classical logic applied in a highly organised manner which the general public sees in the form of space flight, nuclear power and computers, for example. At the same time it will be recalled that there have also been periodic disasters that make national news, some natural others not such as aircraft and rail crashes, nuclear plant failures, submarine losses and many other types due to technical failures or human unreliability. Besides the catastrophic disasters that readily spring to mind, there are also cases of failure, not small in number, that could be traced to design failures, many of these do not make headline news. This does not include cases where a system simply does not meet the required performance specification and the loss is at the lower end of the criticality scale. Increasingly, system reliability is high on the public agenda.

Amongst the most valuable assets of any organisation is that of its professional knowledge base, part of which will be documented and explicit and part of which is dispersed and will reside in the minds of the organisation's staff. The latter is "free" in the sense that it has not been reduced to document form, it is volatile as staff transfer between different organisations. The total knowledge base governs the capacity and proficiency of an organisation to react to technical problems. In practice it would frequently be the case that the induction process for creating relationships would be accomplished by searching the free knowledge base to obtain consensus expert opinion, which may include contracted judgement. Fuzzy logic methods fit naturally and easily into the broader picture of knowledge engineering and more generally into asset management. These factors are to a large extent unrecognised in current engineering practice and offer the potential for a profound change in outlook at the strategic level. This text approaches the issue at a tactical level, but the full possibilities will be apparent.

Fuzzy logic has a much wider and deeper foundation than is implied in this practical text. To learn a new language one can study the grammar or learn by examples. Professional engineers and students with time constraints would usually opt for the latter method and this is the aim in both this and the companion text, *An Introduction to Fuzzy Logic Applications*. The benefits of the new language are; more flexibility and generality in the formulation and solution of problems; non-linear problems are easily encompassed. Also input information is often more fully represented than in conventional treatments and furthermore carried through to the conclusions which consequently have a higher information content. This is a key factor, enabling more reliable decisions to be made by professional engineers and designers.

This work extends and complements that the companion text. It opens up new avenues of applications and new aspects of fuzzy logic. There is a stronger emphasis on the interpretation of the conclusions.

This radical new approach to problems in engineering science and also to professional engineering procedures avoids over representing the precision of information and knowledge and the approximate use of classical logic which is implicit in current practice. Applications in the research literature are fairly sparse and in the professional literature they are practically non-existent. It is anticipated that the absorption of fuzzy logic methods into engineering practice will take time, but it is certain that the advantages and the need to reappraise the current processing of information and formulation of knowledge will prevail and that the methods will become part of the engineer's intellectual tool kit. But in the meantime there is the need for a substantial educational drive to propagate the awareness of the potential of fuzzy logic methods. The way in which this will happen will be through further seminars, short courses and undergraduate or postgraduate courses, all of which require the support of suitable texts. At present there is a very limited selection of texts to serve this purpose. This text is addressed to engineering lecturers, research-ers extending the frontiers of knowledge, professional engineers and designers and also students. A hallmark of fuzzy logic methods is that the cultural gap between researchers and practitioners is not apparent, the linguistic formulation of problems and conclusions is equally coherent to both.

I would like to acknowledge the invaluable help given by Derek Duffett and James Harris in the final stages of compiling the text. Any flaws that remain are mine. Comments on any aspects of the text would be welcome.

John Harris
October 2003

Chapter 1
Comments and Definitions

Most people during the course of their education acquire a knowledge of classical logic, even if it is only of a hazy nature. The hallmark is an acceptance that statements are either "true" or "false": the car will or will not start, the tennis match has or has not been postponed. Such is the basis of classical (Aristotelian) logic (CL). Later, one appreciates that some propositions have an expectation element when they are concerned with future events: the car probably will not start tomorrow morning (based upon experience), the tennis match will probably be postponed if the weather is bad. Such uncertainty is sometimes treated by statistical methods, based upon historical data, though at a less sophisticated level, more often on simple guesswork.

Although CL is a special case of fuzzy logic (FL), its greater familiarity makes it a useful point of departure in this text.

(A guide to fuzzy logic is provided in the companion text, *An Introduction to Fuzzy Logic Applications*. See also Further Reading.)

1.1. Classical Logic Basics

1.1.1. DEFINITIONS

Sets. The basic concept in CL, as in FL, is that of a set. This is conceived as a collection of objects, which may be tangible or intangible, having some common attribute or feature, such as shape, colour, type or use (for example). The common denominator is the categorical characteristic which is the defining feature of the set. A set may comprise, for example, the whole numbers between 0 and 10, another example could be the members of staff of a particular hospital. Sets are denoted in this text by capital letters, thus X.

Subsets. These are sets within sets and have two attributes, one defining the set and one the subset. Subsets are also denoted by capital letters, thus $Y \subset X$, where $\subset$ means subset Y is contained within set X.

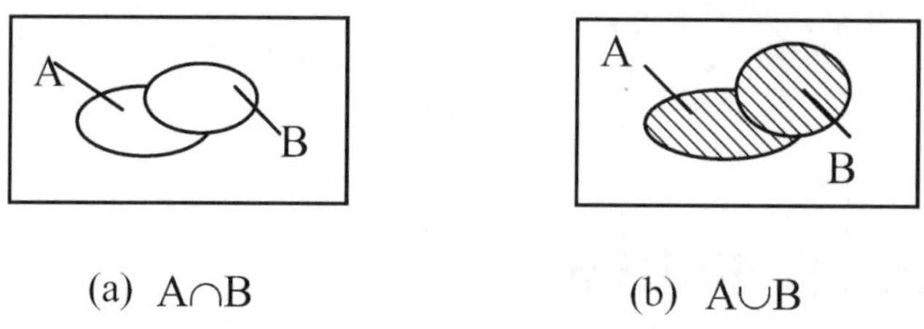

(a) A∩B   (b) A∪B

Figure 1.1. Venn diagrams of classical logic operations. (a) Intersection. (b) Union.

Elements. The individual members of a set are called its elements, and are denoted by lower case letters, thus p. Also, $p \in P$ means that element p is a member of the set P. Elements may be one or more discrete features, or point values of a continuum. The set containing only one element is called the unit set.

If the set is to be specified by listing all the elements, it is written thus; $\{a, b, c\}$, in which the order of the elements within the brackets is immaterial. As an example, $S = \{0, 1\}$ represents the set of electric light switch positions. The switch is either "on" or "off".

Special sets.

(i) Universal set. This comprises all the elements in the population. It is denoted by 1.
(ii) Null set or empty set. This contains no elements. It is denoted by ∅.

Complimentary set. The compliment of set X is set X', where $X + X' = 1$, which is the universal set. It follows that the complement of the universal set is the null set.

Logic Operations. Several CL operations on sets are defined. The intersection AND is denoted by ∩. The union OR is denoted by ∪. The two operations are conveniently illustrated by means of the Venn diagrams, as shown in Figures 1.1(a) and 1.1(b).

In the Venn diagrams in Figure 1.1, the surrounding rectangle represents the universe of all sets in the genre.

The OR operation shown above is the inclusive OR, it must be distinguished from the exclusive OR labelled XOR. This means set A or B, but not both. The XOR operation is not used in this text. Other operations in the literature are NAND and NOR, neither of which are needed here.

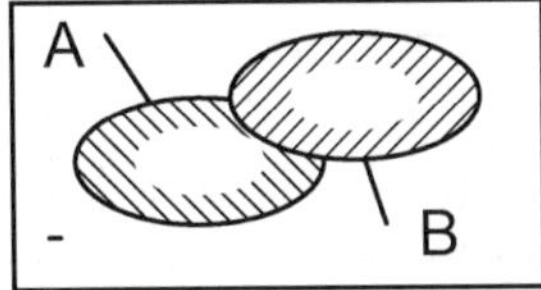

Figure 1.2. Venn diagram showing fuzzy intersection.

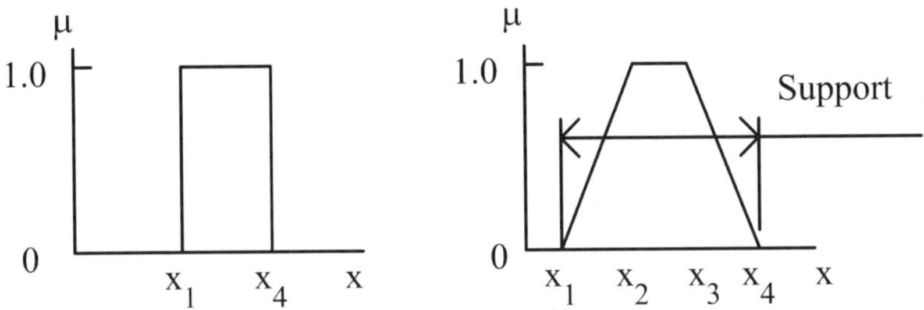

Figure 1.3. Comparison of typical set membership patterns. (a) Classical set. (b) Fuzzy set.

1.2. Fuzzy Logic (FL)

1.2.1. FUZZY SETS AND VENN DIAGRAMS

The sets in CL all have precise boundaries. In FL this requirement is relaxed and therefore the set boundaries in this case are imprecise. The sets on a FL Venn diagram therefore appear as shading, illustrating boundary zones. The imprecise shaded zone defining the boundaries of the sets are the zones in which partial membership, μ $(0 < \mu < 1.0)$ of the sets by the elements is attributed. A fuzzy Venn diagram is illustrated in Figure 1.2

The pattern of membership values of elements of sets is often conveniently displayed on a diagram. Figure 1.3 compares membership functions in the CL and FL cases, where the elements are members of a *continuum*. Values of the elements are set off along the abscissa and membership values of the elements are set off along the ordinate. In Figure 1.3(a), for $0 < x < x_1$, the membership value is zero. For $x_1 < x < x_4$ the membership value is unity, whilst for $x > x_4$ the membership value is again zero. Figure 1.3(a) defines a CL set. In the case of Figure 1.3(b), for $0 < x < x_1$ the membership value is zero, whilst for $x_1 < x < x_2$ it rises from zero to unity. For $x_2 < x < x_3$ the membership value is unity and for $x_3 < x < x_4$ the membership declines again from unity to zero. Beyond x_4 the membership value is again zero.

1.2.2. FUZZY SET SHAPES

The geometry of fuzzy set shapes may take on a variety of forms, but is subject to the requirement that any element must not have more than one membership value of a particular set. Simple possible shapes are illustrated in Figure 1.4. The singleton

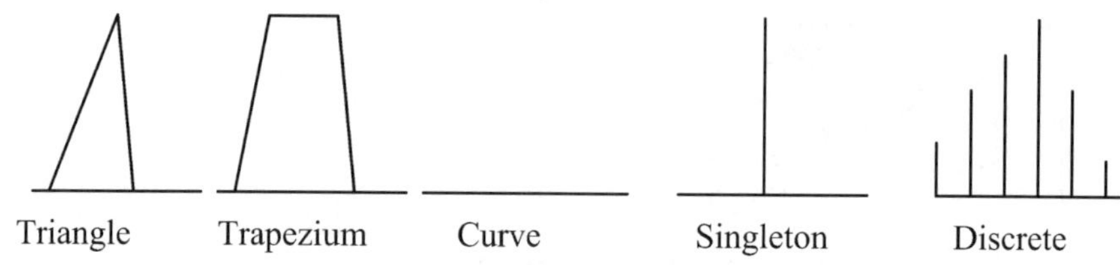

Figure 1.4. Typical fuzzy set membership function shapes.

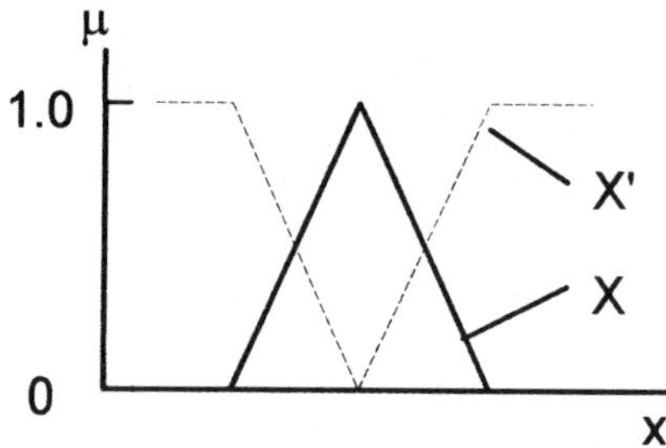

Figure 1.5. A fuzzy set and its complement.

is a special case of a set with only one element, which has unity membership value. Fuzzy sets may comprise continuous or discrete distributions of members. It is sometimes convenient in applications to approximate a continuous distribution of elements by a discrete distribution. The triangular form of membership function is the most common, and will be extensively used in this work.

1.2.3.　THE COMPLEMENTARY FL SET

If X is a fuzzy set, then the complimentary fuzzy set is defined as $X' = 1 - X$. The set and its complement are illustrated in Figure 1.5.

1.2.4.　UNIVERSE OF DISCOURSE (UD)

The elements of a universe are usually distributed amongst a number of sets and in FL they are often members of more than one set. They generally have fractional membership values of the several sets. The fuzzy sets are defined on a universe of discourse of categories of a particular attribute. For example, Figure 1.6 shows sets on a UD of speed in which the categories are varying degrees of "fast".

1.3.　Measurement Scales

Although in engineering practice most elements are described in terms of numerical continua, there are various types of scales that are used. UDs can be described on several different scales, each with it own application. Four types are described below:

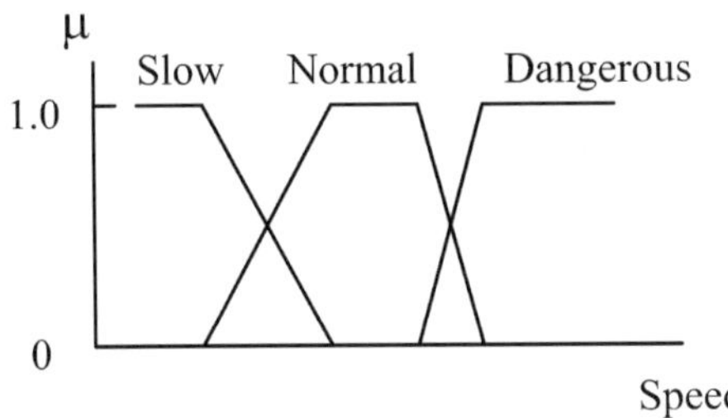

Figure 1.6. The universe of discourse of speed.

(i) Ordinal scales. The elements on this scale are placed in some sort of order of progression, for example, the stations on a railway line or items in a programme.

(ii) Nominal scale. The elements represent discrete categories or conceptual objects, for example, types of screw thread or types of non-ferrous metal.

(iii) Interval scale. Units or intervals of a given size marked out in sequence, for example, the intervals on a clock face.

(iv) Ratio scale. An interval scale with a zero point at some arbitrary or agreed value, for example, a tape measure division.

This offers a range of scales suitable for different applications. The essential feature of a scale is that it reflects a common theme amongst the elements which defines the UD.

1.4. Propositions

Logic processes are conducted in the form of propositions, the simplest of which comprises an antecedent (premise) and a conclusion, IF A THEN B, e.g., if that is a litre of water then it will weigh one kilogram (at sea level). CL asserts that the membership value is 1 or 0.

Consider a FL proposition: If the brake pedal is pressed, then the car will stop. Now there are various levels of brake pressure that may be applied and there are various levels of stopping from an emergency stop to gradual slowing down. Thus there are various categories of stopping and a CL conclusion is not possible. The resolution of this type of problem is the theme of the remainder of this text.

Compound propositions can take several different forms. One of the more important and frequently used forms is: IF A AND B THEN C, for example, if the distance of a hotel along a road is possibly between one and five kilometers and the distance to a garage is possibly between three and seven kilometers (both $+ve$) then the distance between them is somewhere between three and five miles. This case is illustrated in Figure 1.7.

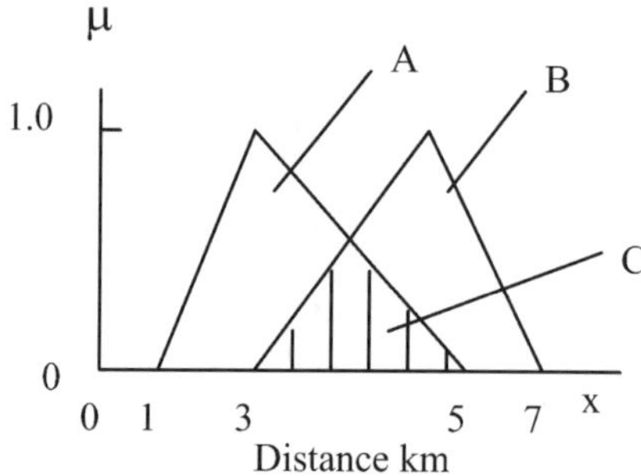

Figure 1.7. Illustration of fuzzy intersection.

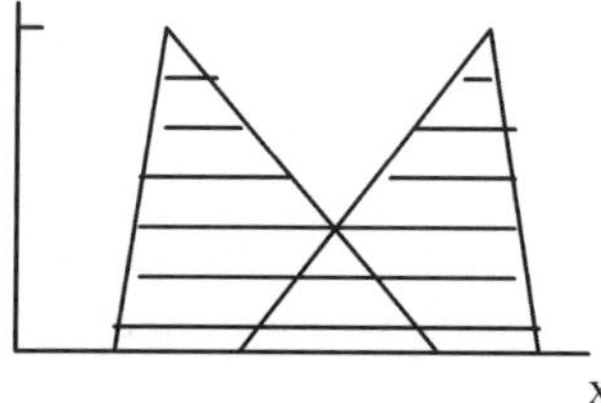

Figure 1.8. Illustration of fuzzy union.

Another important and often used case of a compound proposition is expressed as: IF A OR B THEN C. For example, if the wind is strong or very strong then it is unsafe for novice sailors to race. This case is illustrated in Figure 1.8.

1.5. Notation

In this work sets and subsets are labelled with capital letters. The elements of a set are written in lower case letters and contained in brackets { }. A fuzzy number is expressed as a fuzzy set. A fuzzy number may be represented in discrete or continuous form. The discrete form is of the type

$$A = [\mu_1//a_1 + \mu_2//a_2 + \mu_3//a_3 + \cdots]. \tag{1.1}$$

The elements with their membership values (μ_i) are enclosed in square brackets [ ]. The membership value is separated from its corresponding element (a_i) by the symbol //; this does not denote devision. The + sign within the [ ] brackets denotes continuation, it does not denote addition.

The *continuum* form of fuzzy number is expressed as

$$X = \int \mu(x)//x, \tag{1.2}$$

where $\int$ does not mean "integral of ...", but "continuous distribution of ...".

In the solution of problems it is often useful to represent fuzzy sets as piecewise continuous distributions. Discrete values are also used to represent continuous dis-

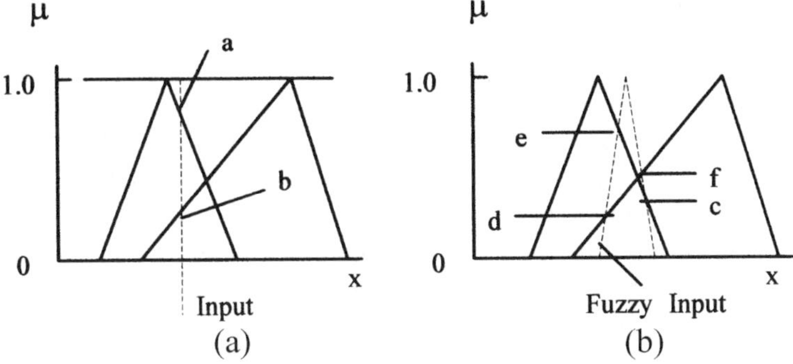

Figure 1.9. Assigning input data membership values. (a) Singleton input. (b) More general fuzzy input.

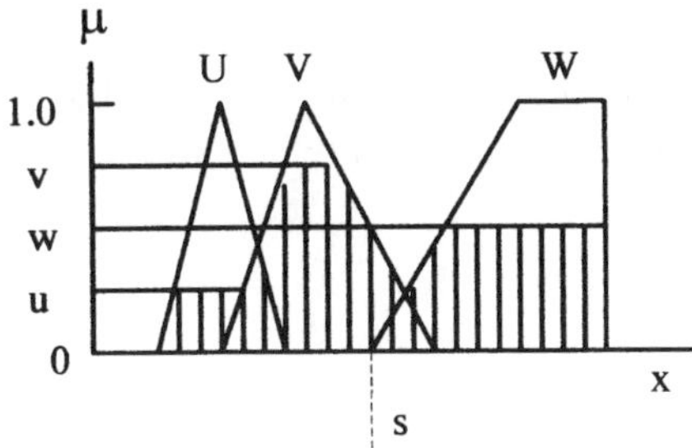

Figure 1.10. The union of several fuzzy subsets.

tributions. The volume of work in manipulating discrete sets may be reduced by considering principal sets, as defined in the Appendix.

1.6. Fuzzification and Defuzzification

1.6.1. FUZZIFICATION

The operations in FL are performed in terms of fuzzy sets. In practice, the input data may also be in terms of fuzzy sets or a singleton (single element with a membership value of unity), which is infact a special type of fuzzy set. The input data needs to be assigned membership values of one or more fuzzy sets into which the UD has been partitioned. The membership values are found from the intersections of the data sets with the fuzzy sets of the UD. Figure 1.9(b) illustrates the graphical method of finding membership values in the case of a singleton (Figure 1.9(a)), and the more general fuzzy input (Figure 1.9(b)).

For the singleton in Figure 1.9(a), there are two intercepts, i.e., at a and b, which determine the membership values. Whilst for the fuzzy input in Figure 1.9(b) there are four intercepts at c, d, e and f which determine the membership values.

1.6.2. DEFUZZIFICATION

This means the reduction of the fuzzy set or subset to a singleton. The fuzzy set is usually the union of several subsets representing the conclusion of a fuzzy proposition. Normally, a fuzzy set cannot be represented by a singleton, therefore defuzzification can only be undertaken with the loss of information. The union of several subnormal (no membership value equal to unity) fuzzy subsets is illustrated in Figure 1.10 and s is the single element on the UD which is deemed to represent the union of the fuzzy subsets. Such a representation discards the span of the conclusion and the membership values of the subsets. But for calculations in design (for example) a specific value is required and provides the motivation for defuzzifying, but it is important not to loose sight of the whole solution.

There are several ways of finding a representative number. Two common ways are outlined below.

(i) Centroid method. This is probably the most frequently used method and as the name suggests, it involves finding the position of the centre of area of the subsets on the abscissa (s).

$$\text{Continuous distribution:} \quad s = \int_{x=0}^{x=\infty} x \, da \Big/ \int da, \tag{1.3}$$

$$\text{Discrete distribution:} \quad s = \sum_{i=1}^{i=n} x_i \delta A_i \Big/ \sum \delta A_i. \tag{1.4}$$

(ii) Weighted abscissa method. This is evaluated by taking the sum of the normalised weighting of each of the set principal values, $x_i(\max)$

$$s = \sum_{i=1}^{i=n} \mu_i x_i(\max) \Big/ \sum \mu_i. \tag{1.5}$$

In the trapezoidal shape of fuzzy set, it is the mid-support value that is used for x_i. The centroid and weighted abscissa methods generally give somewhat different values of the defuzzified representative number. It may be observed that given a representative number, it is generally not possible to recover the original fuzzy subsets.

1.7. Equivalent (Triangular) Fuzzy Number (EFN)

Given the union of two fuzzy sets, as illustrated in Figure 1.11, it is possible to find a symmetrical triangular fuzzy number that would produce the same fuzzy subsets by intersection with the partitioning fuzzy sets on the UD. This is called the Equivalent Fuzzy Number (EFN). There is less loss of information with this representation than with the defuzzified number. Given an EFN, it is possible to

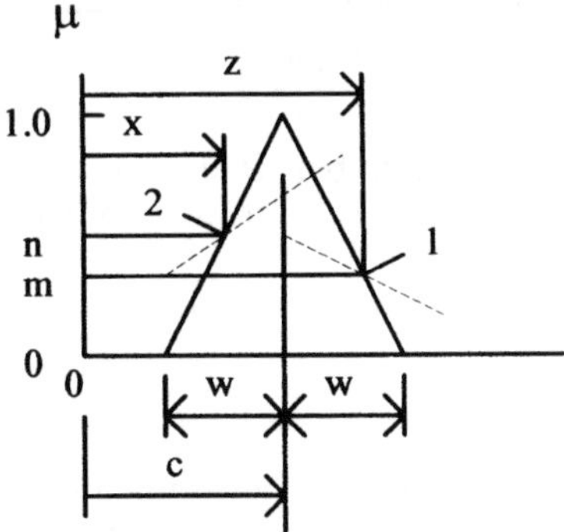

Figure 1.11. An Illustration of the equivalent fuzzy number.

recover two pairs of fuzzy subsets, one of which is the original fuzzy subset, but generally not with a defuzzified number.

Figure 1.11 shows a symmetrical EFN in which points 1 and 2 represent points of intersection of the EFN with the two different fuzzy sets on the UD. The membership values of the intersection points 1 and 2 are m and n, respectively, therefore from the membership functions of the fuzzy sets the values of z and x at the intersections may be obtained. Now by similar triangles,

$$(c - x)/(1 - n) = (c + w - z)/m = w/1.$$

Hence,

$$c + w - z = mw. \tag{1.6}$$

Therefore,

$$c - z = w(m - 1). \tag{1.7}$$

Also,

$$c - x = w(1 - n), \tag{1.8}$$

Subtracting Equation (1.7) from Equation (1.8),

$$z - x = w(2 - n - m).$$

Hence,

$$w = (z - x)/(2 - m - n). \tag{1.9}$$

Also from Equation (1.7),

$$c = x + w(1 - n). \tag{1.10}$$

Equations (1.9) and (1.10) determine the principal value (c) and the support of the EFN ($2w$). At the cross-over point $z = x$ and hence $w = 0$, the fuzzy number degenerates to a singleton. The principal value of the EFN may be viewed as a representative number of the union of the two fuzzy subsets from which the EFN was formed.

For symmetrical sets and a symmetrical fuzzy number, if $n + m = 1$, then $w = y - x$. This is generally only true if $y - x = 0$, i.e., if $w = 0$. The fuzzy number then degenerates to a singleton.

1.8. Probability and Identity

The first impressions of fuzzy logic analysis is that it is statistics with another label and adds nothing to the clarification and solution of non-deterministic problems. This is a fundamental misunderstanding of the nature of fuzzy logic. To set things in their proper order, it may be stated that classical two-valued logic is a special case of fuzzy logic and statistics is based upon CL. The source of the confusion between the two lies in the use of the word "probably", which in many cases connotes the frequency of an event in a given static population, but can have different interpretations. However, fuzzy logic is not concerned with frequency but with the identity of an individual conceptual object, be it real or non-material, and which may or may not be a member of a population (which may not be static). It is a fact that both probability theory and fuzzy logic are concerned with uncertainties, but they are uncertainties of an entirely different nature.

Consider a given bag of mixed black and white balls, the probability of withdrawing a black ball or a white ball is simply found, given the total number of black and white balls contained in the bag. The balls are unmarked and one black (or white) ball is indistinguishable from another. This is a probability problem. Suppose now that the balls are various shades of grey ranging from pure white to pure black. In this case each ball would need to be examined and assigned a membership value on the scale 0–1.0 of the black category and of the white. Each ball would generally belong partly to the one category and partly to the other. This is a fuzzy logic problem, not soluble by probability theory, it is a matter of identity. Engineering practice abounds in problems of uncertainty of the identity kind. In many ways engineering logic is fuzzy logic.

Chapter 2
Fuzzy Geometry

Geometric considerations arise in the context of engineering studies in many ways, especially in the mechanical and civil engineering and related fields, also in architecture. In many, but not all cases, geometric aspects arise at the design stage of artefacts and systems, ranging from large one-of-a-kind designs such as bridges, ships or process plant, to mass produced domestic white goods, furniture and motor cars, for example. Visual appeal, ease of assembly, safety, performance and interchangeability are all important functions which have a greater or lesser dependence on geometric factors.

Geometry as an academic study is normally concerned with theoretical forms that have idealised shapes and exact dimensions. But these have no precise realisation in nature or in industrial artefacts. The latter are produced to dimensions that have allowable variations from the theoretically exact values, the variations are as generous as possible (high precision is expensive). These allowances are formalised as tolerances, which means imposed limits to the acceptable variations of the dimensions, acknowledging that a complete set of theoretically exact dimensions is improbable. Thermal, wear and load effects also cause dimensional changes that are additive to the manufacturing errors. There are also measurement errors.

The solution of several cases of exact geometric relations, but with fuzzy input data is described, which introduces the application of the extension principle. The solution of a problem with fuzzy geometric relationships and fuzzy input data is also given. Finally, the blending of planar curves and the fairing of sections is described.

2.1. Linear Measurement

The accuracy of the linear distance or dimension between two points depends upon the precision of the identity of the location of the reference points (assuming perfect accuracy of the measuring instruments). This is a matter of tacit agreement, reached in practice when the value-at-risk due to differences of opinion on the locations are insignificant. In manufactured goods, if the value-at-risk due to variations in a linear dimension is significant, then dimensional limits would be imposed. If the problem related to surface texture then similar limits would be imposed. In the

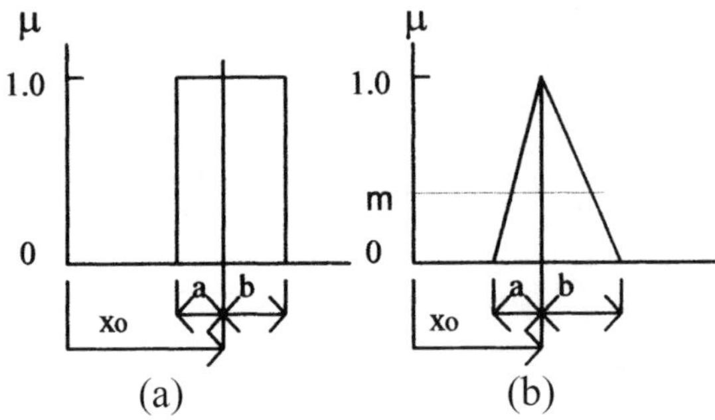

Figure 2.1. Illustrations of dimensional tolerances. (a) Conventional CL type. (b) General FL type.

former case, the limits on a dimension are often the bilinear type $(x_0 - a, x_0 + b)$ where x_0 is the precise dimension, which has a lower limit of $(x_0 - a)$ and an upper limit of $(x_0 + b)$. An acceptable dimension is one with the value lying between these limits. This is a CL type of criterion, as shown in Figure 2.1(a), which defines the set of acceptable values of x on the UD. The tolerance in this case is $(a + b)$.

If the tolerance zone is now reinterpreted as a fuzzy set, a more general and flexible specification may be achieved which can accommodate a range of quality (precision) criteria within the same basic envelope. $(a + b)$ defines the support of the fuzzy tolerance set.

A series of quality levels may be formed within the envelope in addition to the basic level with a tolerance of $(a + b)$. For example, a higher quality level could be specified as "better than m", referring to Figure 2.1(b). Thus, several quality levels could be established with different values of μ. Other applications, such as the example below, utilise a similar concept.

The same principle could clearly be extended to other applications, including non-geometric cases, such as chemical composition. In the case of natural objects, variability and uncertainty of size and shape are the rule, nature is exceedingly rich in its tolerance to variations of natural form. Fuzzy sets are much more compatible with reality than are classical sets.

EXAMPLE 2.1. The distance between opposite points on the banks of a river have been measured at two places, namely A and B as shown in Figure Ex 2.1. The estimates of the distances in metres are given as discrete fuzzy sets in which the membership values indicate the most likely distance.

$$A = [0//7.5 + 0.6//7.8 + 1.0//8.0 + 0.5//9.2 + 0//9.4],$$

$$B = [0//11.7 + 0.7//11.8 + 1.0//12.0 + 0.3//12.3 + 0//12.5].$$

Find the difference between these two measurements, assuming that the measurement errors are negligible.

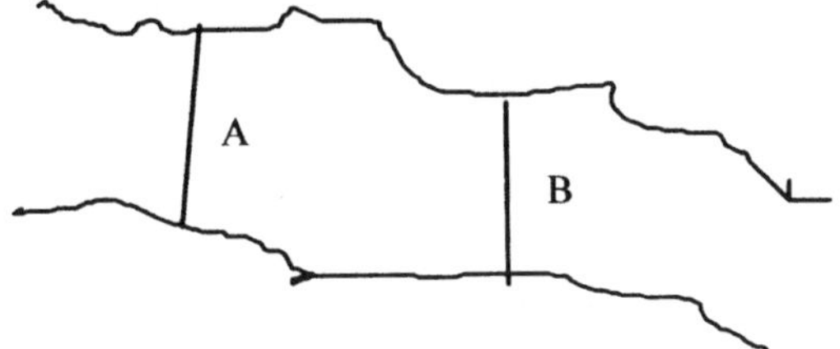

Figure Ex 2.1. Distance of opposite river banks.

Solution. The difference (D) between the two fuzzy measurements, A and B, is given by the Cartesian product of the membership values and the arithmetic difference of the associated measurements. Thus,

$$D = B \times A//(b - a).$$

Now,

$$B \times A = \min[(\mu_A, \mu_B)]$$

$$= [0, 0.7, 1.0, 0.3, 0]x[0, 0.6, 1.0, 0.5, 0]$$

$$= (0, 0)(0, 0.6)(0, 1.0)(0, 0.5)(0, 0) = (0, 0, 0, 0, 0), \text{ etc.}$$

The complete arrays of membership and arithmetic values are

		7.5	7.8	8.0	9.2	9.4	
	11.7	0	0	0	0	**0**	
	11.8	0	0.6	0.7	**0.5**	0	
$B \times A//(b-a) =$	12.0	0	0.6	**1.0**	0.5	0	//
	12.3	0	**0.3**	0.3	0.3	0	
	12.5	0	0	0	0	0	

4.2	3.9	3.7	2.5	**2.3**
4.3	4.0	3.8	**2.6**	2.4
4.5	4.2	**4.0**	2.8	2.6
4.8	**4.5**	4.3	3.1	2.9
5.0	4.7	4.5	3.2	3.1

The principal set values are shown in bold type in the above arrays and are displayed as a piecewise continuous set in Figure Ex 2.2. The discrete form of the set is,

$$D = [0//2.3 + 0.5//2.6 + 1.0//4.0 + 0.3//4.5 + 0//5.0]m$$

Defuzzifying the principal set using the weighted abscissa method (Equation 1.5)), gives

$$\text{Def } D = (0.3 * 4.5 + 1.0 * 4.0 + 0.5 * 2.6)/(0.3 + 1.0 + 0.5)$$

$$= 3.694 \text{ m.}$$

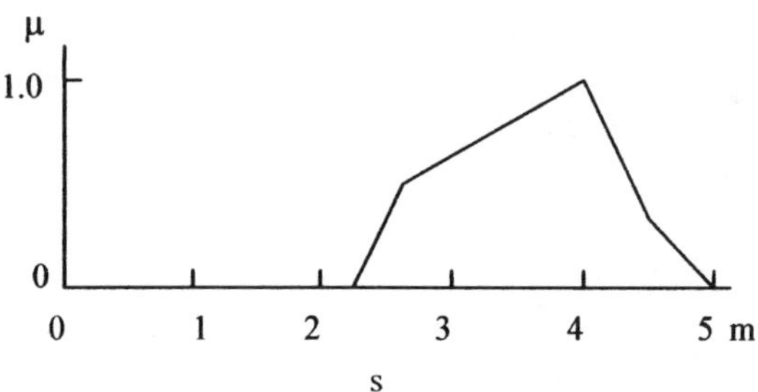

Figure Ex 2.2. The fuzzy difference $(B - A)$ represented as a piecewise continuous function.

Using average values of A and B, the difference is 3.68 m. But neither this nor the defuzzified value reflect the full information in the fuzzy value of D, which spans the range, 2.3 m to 5.0 m, with a value of 4.0 m at the maximum membership level. It is not suggested that Def D represents a more correct value than the average of 3.68 m. Fuzzy D represents the correct solution.

If the objective was, for example, to span the river with a bridge, then the design team would need to select the membership value on safety grounds.

2.2. Fuzzy Areas

An area may be bounded by a fuzzy perimeter which makes the calculation of the area uncertain. A common approach to such a problem would be to define a mean perimeter and then calculate a precise area based upon this. Consonant with the previous discussion, such a methodology discards useful information about the range of the values of area implied within the fuzzy perimeter data. There are various categories of shape treated by different methods:

(i)   Simple geometric shapes whose areas are described by known deterministic formulae.
(ii) More complex shapes requiring treatment from first principles.
(iii) Forms comprising blended simple shapes.

Types (i) and (ii) are soluble by the application of the extension principle. Type (iii) can be solved by combining piecewise solutions of component shapes using membership functions. A simple example of type (i) is that of the area of a fuzzy rectangle, as described below. Type (iii) problems are considered later in this chapter. Type (ii) problems may be solved by the application of the extension principle, they are not treated here.

2.3. Fuzzy Rectangle

Consider a plane rectangle in which the dimensions of the sides are represented by fuzzy sets. Let the fuzzy dimensions of the rectangle, B and C, be given by three-term discrete sets,

$$B = [\mu_{B1}//b_1 + \mu_{B2}//b_2 + \mu_{B3}//b_3] \tag{2.1}$$

and

$$C = [\mu_{C1}//c_1 + \mu_{C2}//c_2 + \mu_{C3}//c_3]. \tag{2.2}$$

The fuzzy area (A) of the rectangle is then found by the Cartesian product of the membership functions and the algebraic product of the associated elements,

$$A = B \times C//(b * c). \tag{2.3}$$

Now,

$$B \times C = \min[(\mu_B, \mu_C)], \tag{2.4}$$

where

$$\min[(\mu_B, \mu_C)] = \min[(\mu_{B1}, \mu_{C1}), (\mu_{B1}, \mu_{C2}) \ldots \text{ etc.} \tag{2.5}$$

The solution may then be expressed in terms of two arrays, one of which comprises membership values and the other the algebraic products.

EXAMPLE 2.2. Let the sides of a rectangle be given in discrete fuzzy form by

$$B = [0//0 + 0.5//0.5 + 1.0//1.0 + 0.5//1.5 + 0//2.0] \quad \text{and}$$

$$C = [0//1.0 + 0.5//1.5 + 1.0//2.0 + 0.5//2.5 + 0//3.0].$$

Find the area of the rectangle.

Solution. The sets representing the rectangular sides are illustrated as piecewise continuous sets in Figure Ex 2.3.

The membership value array obtained by performing the Cartesian product, formulae (2.3, 2.4) and 2.5) on the given discrete fuzzy sets B and C, together with the associated arithmetic products array, are tabulated below. The principal discrete set numbers of the area are shown in bold type. Figure Ex 2.4 shows these results as a piecewise continuous set.

$$A = \begin{vmatrix} \mathbf{0} & 0 & 0 & 0 & 0 \\ 0 & \mathbf{0.5} & 0.5 & 0.5 & 0 \\ 0 & 0.5 & \mathbf{1.0} & 0.5 & 0 \\ 0 & 0.5 & 0.5 & \mathbf{0.5} & 0 \\ 0 & 0 & 0 & 0 & \mathbf{0} \end{vmatrix} // \begin{vmatrix} \mathbf{0} & 0 & 0 & 0 & 0.0 \\ 05 & \mathbf{0.75} & 1.0 & 1.25 & 1.5 \\ 1.0 & 1.5 & \mathbf{2.0} & 2.5 & 3.0 \\ 1.5 & 2.225 & 3.0 & \mathbf{3.75} & 4.5 \\ 2.0 & 3.0 & 4.0 & 5.0 & \mathbf{6.0} \end{vmatrix}.$$

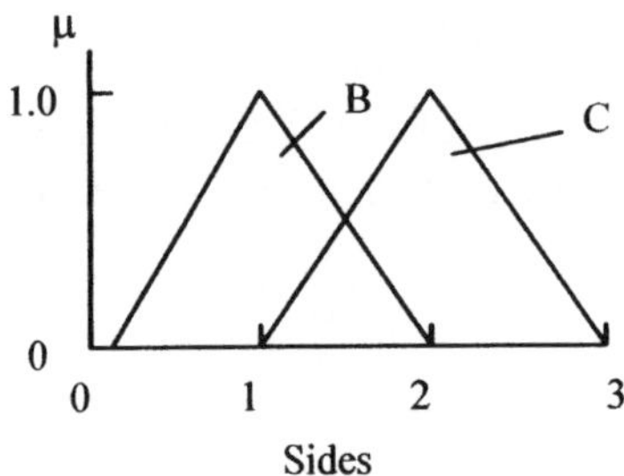

Figure Ex 2.3. The sides of the rectangle as piecewise continuous sets.

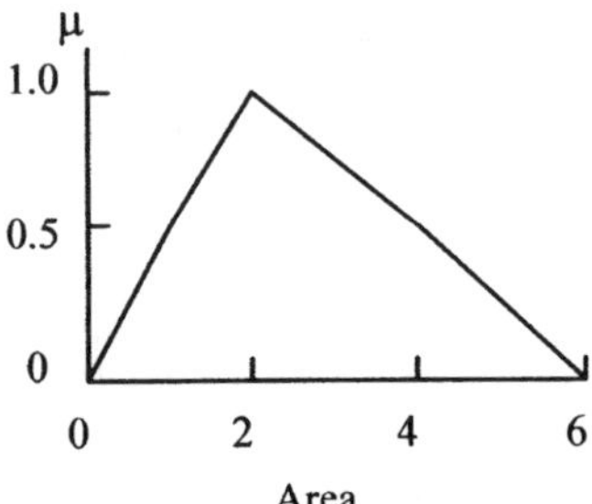

Figure Ex 2.4. The principal set values of the area of the rectangle as a piecewise continuous set.

The principal set of the area is

$$A = [0//0 + 0.5//0.75 + 1.0//2.0 + 0.5//3.75 + 0//6.0].$$

Defuzzifying the principal fuzzy set using the weighted abscissa function method gives

$$\text{Def } A = \Sigma \mu_i a_i / \Sigma \mu_i$$
$$= (0.5 * 0.75 + 1.0 * 2.0 + 0.5 * 3.75)/(0.75 + 1.0 + 0.5).$$

Therefore,

$$\text{Def } A = 1.89.$$

The area is not a precise value, Def A indicates a tendency.

The numerical average of the side B is 0.8333 and of side C is 2.000 and the product of the average values gives an area of 1.667. It will be noted that the symmetrical sets of the sides of the rectangle produces an asymmetric fuzzy set area with a support (span) of 6 units.

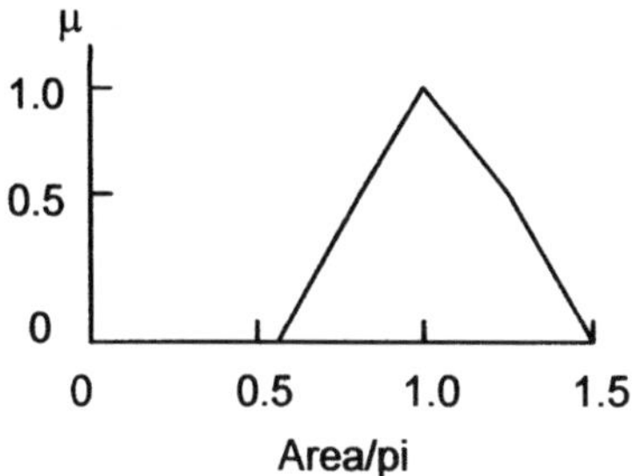

Figure Ex 2.5. The area of a circle as a piecewise continuous function.

2.4. Fuzzy Circle

Another simple geometric shape is represented by a circle. In the previous section
the area of a rectangle was found by Cartesian product, in this case the extension
principle will be applied to find the area.

Let r be the radius of a circle in discrete fuzzy set form. The area (a) is given
by the conventional formula

$$a = \pi r^2. \tag{2.6}$$

Let the radius of the fuzzy circle be given in discrete form by

$$R = [\mu_1//r_1 + \mu_2//r_2 + \mu_3//r_3 + \mu_4//r_4 + \mu_5//r_5]. \tag{2.7}$$

Then by the extension principle the corresponding area (A) is given by

$$A = [\mu_1//\pi r_1^2 + \mu_2//\pi r_2^2 + \mu_3//\pi r_3^2 + \mu_4//\pi r_4^2 + \mu_5//\pi r_5^2], \tag{2.8}$$

or

$$A = [\mu_1//r_1^2 + \mu_2//r_2^2 + \mu_3//e_3^2 + \mu_4//r_4^2 + \mu_5//r_5^2]\pi, \tag{2.9}$$

where it is understood that the π factor applies to each term in Equation (2.9).

EXAMPLE 2.3. A fuzzy circle has a radius given by

$$R = [0//0.8 + 0.5//0.9 + 1.0//1.0 + 0.5//1.1 + 0//1.2].$$

From Equation (2.9) the area of the fuzzy circle is

$$\begin{aligned}
A &= [0//0.8 + 0.5//0.9 + 1.0//1.0 + 0.5//1.1 + 0//1.2]\pi \\
&= [0//0.64 + 0.5//0.81 + 1.0//1.0 + 0.5//1.21 + 0//1.4]\pi.
\end{aligned}$$

This is a slightly asymmetrical set. It is represented as a piecewise continuous set
in Figure Ex 2.5.

The same result may be obtained by the Cartesian product method, $A = \pi(R \times R)$. This yields an array in which the principal set provides an identical solution.

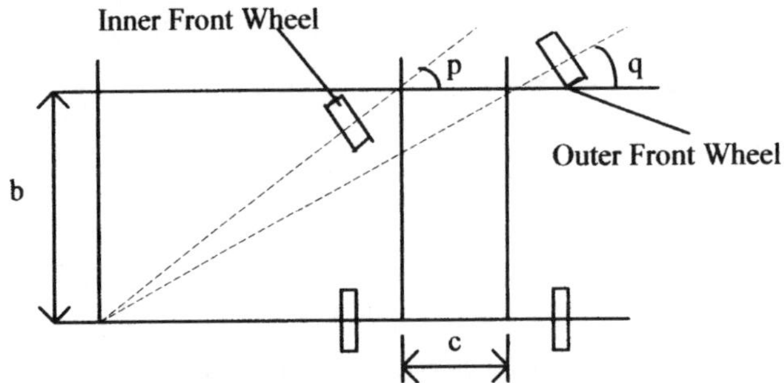

Figure Ex 2.6. Vehicle steering geometry.

The above example demonstrates the use of a simple exact theoretical relationship with fuzzy data. A more practical type of problem, again using the Cartesian product principle, is described below.

EXAMPLE 2.4. In a four-wheeled vehicle with front wheel steering, the condition for the theoretically exact angular relationship between the front wheels is given by

$$\cot p - \cot q = c/b,$$

where b, c, p and q are defined in Figure Ex 2.6.

Manufacturing and assembly tolerances allow for variations in the ratio c/b. For this vehicle the ratio is specified as

$$C/B = [0//0.437 + 0.5//0.499 + 1.0//0.460 + 0.5//0.472 + 0//0.483].$$

In a worn and poorly assembled condition, the outer wheel in a cornering manoeuvre has an angle (q) given by

$$Q = [0//27.6 + 0.5//28.8 + 1.0//30.0 + 0.5//31.2 + 0//32.4] \text{ deg}.$$

What would be the corresponding angle (p) of the inside front wheel?

Solution. If

$$Q = [0//27.6 + 0.5//28.8 + 1.0//30.0 + 0.5//31.2 + 0//32.4] \text{ deg},$$

then

$$\cot Q = [0//1.913 + 0.5//1.819 + 1.0//1.732 + 0.5//1.651 + 0//1.576].$$

Also if

$$C/B = [0//0.437 + 0.5//0.449 + 1.0//0.460 + 0.5//0.472 + 0//0.483],$$

the Cartesian product of $(\cot Q - C/B)$ is

$$\cot Q - C/B = \begin{vmatrix} \mathbf{0} & 0 & 0 & 0 & 0 \\ 0 & \mathbf{0.5} & 0.5 & 0.5 & 0 \\ 0 & 0.5 & \mathbf{1.0} & 0.5 & 0 \\ 0 & 0.5 & 0.5 & \mathbf{0.5} & 0 \\ 0 & 0 & 0 & 0 & \mathbf{0} \end{vmatrix} \Bigg/\Bigg/ \begin{vmatrix} \mathbf{1.476} & 1.464 & 1.453 & 1.441 & 1.430 \\ 1.382 & \mathbf{1.370} & 1.359 & 1.347 & 1.336 \\ 1.295 & 1.283 & \mathbf{1.272} & 1.260 & 1.249 \\ 1.214 & 1.202 & 1.191 & \mathbf{1.179} & 1.168 \\ 1.139 & 1.111 & 1.116 & 1.104 & \mathbf{1.093} \end{vmatrix}.$$

The principal values in the above arrays are printed in bold type. The discrete principal set for $\cot P$ is,

$$\cot P = [0//1.093 + 0.5//1.179 + 1.0//1.272 + 0.5//1.370 + 0//1.479].$$

Therefore,

$$P = [0//34.12 + 0.5//36.13 + 1.0//38.17 + 0.5//40.30 + 0//42.46] \text{ deg.}$$

The spread of the angle q is 4.8 deg, but the spread of the angle p is 8.34 deg.

2.5. Incomplete Restraint

The geometric principles of mechanical systems is treated in the standard theory of machines texts. The nature of the constraints and connectivity between the adjacent bodies is fundamental to the analysis. This is determined by the type of constraining surfaces between them, which are called kinematic elements. If one element of a kinematic pair is fixed in position, the other is constrained to move in the manner allowed by the geometry of the surfaces. Relative motion between adjacent bodies is determined by the following factors:

(i) The possible degrees of freedom of one kinematic element relative to the other.

(ii) The necessary and sufficient conditions of restraint.

It may be shown that any six independent kinematic conditions will suffice to define the position of a body at successive time intervals and thus the locus of the motion of any point on it. But only three independent conditions are necessary if the motion is coplanar. Constraints may be positive, such as in the case of a bearing shaft supported in a journal, or there may be force closure, for example, the spring force applied to a cam follower.

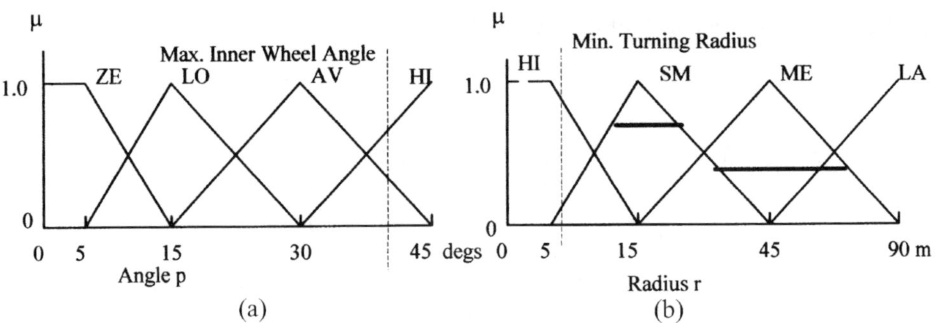

Figure Ex 2.7. Partitioning of the p and r universes of discourse. (a) Inner wheel angle. (b) Turning circle radius.

There is frequently an exact theoretical relationship between the motion of points in constrained adjacent bodies in mechanical systems. In practice however imperfections or errors in the relationships arise due to any one or more of: design clearances and tolerances, wear, deflections under load, inertia forces, thermal expansion or damage. Where the restraint is not positive, for example, if it is by friction, then the relationship may be of an empirical nature. An example of an inexact relationship is given below.

EXAMPLE 2.5. In the previous example, the general relationship between the inner front wheel angle (p) and the turning circle radius (r) measured from the centre of the circle to a point mid-way between the rear wheels, is described by the (P, R) relationship

P	ZE	LO	AV	HI
R	LA	ME	SM	NI

where ZE = Zero, LO = Low, AV = Average, HI = High, NI = Nil, SM = Small medium, ME = Medium, LA = Large.

Possible partitioning of the p and r UDs is shown in Figure Ex 2.7.

Estimate the turning circle for which the inner front wheel angle is expressed as a fuzzy set by

$$P = [0//23 + 0.5//24 + 1.0//25 + 0.5//26 + 0//27] \text{ deg.}$$

Solution. Membership functions for the partitioning in Figure Ex 2.7(a) are tabulated below:

	$0 \leq p \leq 5$	$5 \leq p \leq 15$	$15 \leq p \leq 30$	$30 \leq p \leq 45$ deg
μ_{ZE}	1.0	$1.5 - p/10$		
μ_{LO}		$(p-5)/10$	$2.0 - p/15$	
μ_{AV}			$(p-15)/15$	$3.0 - p/15$
μ_{HI}				$(p-3-)/15$

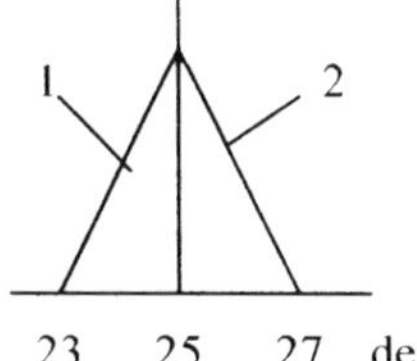

Figure Ex 2.8. The inner front wheel angle as a piecewise continuous set.

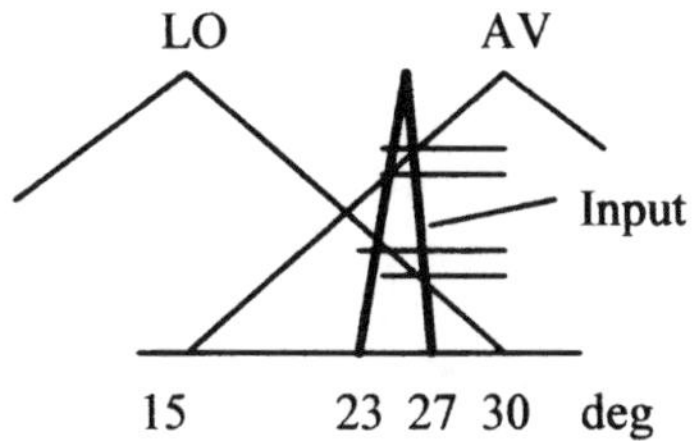

Figure Ex 2.9. The inner front wheel angle intersections.

Turning circle radius membership functions, Figure Ex 2.7(b), are shown below:

	$0 \le r \le 5$	$5 \le r \le 15$	$15 \le r \le 45$	$45 \le r \le 90$ m
μ_{HI}	1.0	$1.5 - r/10$		
μ_{SM}		$(r-5)/10$	$2.0 - p/15$	
μ_{ME}			$(r-15)/30$	$2.0 - r/45$
μ_{LA}				$(r-45)/45$

The input (p) is given as a discrete set,

$$P = [0//23 + 0.5//24 + 1.0//25 + 0.5//26 + 0//27] \text{ deg.}$$

This is illustrated as a piecewise continuous set in Figure Ex 2.8.

For the above representation of P, the membership functions are

$$\mu_i = (p - 23)/2 \quad \text{and} \quad \mu_2 = 13.5 - p/2.$$

The intersection of the above membership functions with the wheel angle membership functions provides two angular values on the LO set and two on the AV set. These are: LO = 23.82 deg; AV = 24.23 deg; LO = 26.54 deg; and AV = 25.59 deg. The intersections are illustrated in Figure Ex 2.9.

Taking the lowest intercepts on the LO and AV sets, the corresponding membership values are

$$\mu_{LO} = 2.0 - p/15 = 2.0 - 26.54/15 = 0.2507,$$

$$\mu_{AV} = (p - 15)/15 = (24.23 - 15)/15 = 0.6153.$$

The FL proposition is

IF P	THEN R	MIN.	CONSEQUENCE
AV	SM	0.6153	0.6153SM
LO	ME	0.2307	0.2307ME

The conclusion on R is the union of the above consequences

$$R_{\mathrm{L}} = 0.6153\mathrm{SM} \cup 0.2307\mathrm{ME}.$$

With some loss of information, a defuzzified value of R may be found by applying the weighted abscissa value method,

$$\mathrm{Def}\, R_{\mathrm{L}} = (0.6153 * 15 + 0.2307 * 45)/(0.6153 + 0.2307)$$

$$= 23.18 \text{ m}.$$

The average value of the fuzzy sets principal values is 30 m.

Now considering the highest intercepts on the LO and AV sets, the corresponding membership values are: LO = 23.82 deg and AV = 25.59 deg. The corresponding membership values are = 0.4120 and = 0.7060.

The FL proposition is

$$
\begin{array}{llll}
\text{AV} & \text{SM} & 0.7060 & 0.7060\text{SM} \\
\text{LO} & \text{ME} & 0.4120 & 0.4120\text{ME}.
\end{array}
$$

The union of the above consequences provides another turning circle radius

$$R_{\mathrm{H}} = 0.7060\mathrm{SM} \cup 0.4120\mathrm{ME}.$$

This conclusion is illustrated in Figure Ex 2.7b.

The defuzzified value of this set is Def $R_{\mathrm{H}} = 26.06$ m. This consequence subsumes the previous expression. The emphasis in both cases is towards the SM set for the full range of axle loads and road surface conditions. More detailed and differentiated data derived from further experimental investigations would enable a more precise solution.

2.6. Blending Coplanar Curves

It often happens in geometric design that two geometric features require to be blended or faired together, for example, when two different sections of a structural member are to be joined. If the resulting feature is free from constraints, other than that of visual appeal, the designer has freedom in the choice of outline. In other cases it may be desired to simulate a given feature, for example, the shape of a tyre-road contact in road research. Such contours may often be represented by patching together two or more simple geometric curves with transitions based upon the concepts of FL rather than by creating mathematical expressions. The method may be exemplified by considering two coplanar circular arcs as shown in Figure 2.2, in which the symbols are defined.

By inspection of Figure 2.2, it may be noted that for the upper circular arc

$$r \sin w = c \sin u - b \tag{2.10}$$

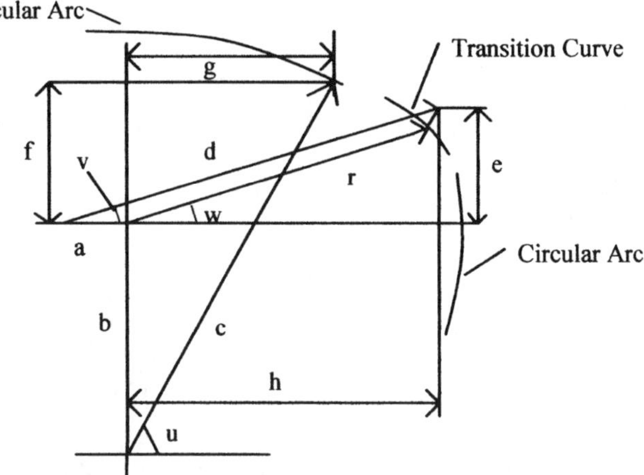

Figure 2.2. Blending two coplanar circular arcs.

and

$$r \cos w = c \cos u. \tag{2.11}$$

Also, for the other circular arc

$$r \sin w = d \sin v \tag{2.12}$$

and

$$r \cos w = d \cos v - a, \tag{2.13}$$

where $r = r(w)$.

Dividing Equation (2.10) by Equation (2.11) and Equation (2.13) by Equation (2.12) yields

$$\tan w = (c \sin u - b)/c \cos u \tag{2.14}$$

and

$$\cot w = (d \cos v - a)/d \sin v. \tag{2.15}$$

Let

$$\tan w = t; \quad \cot w = 1/t; \quad \sin u = \alpha \quad \sin v = \beta.$$

Therefore, $t = (c\alpha - b)/c(1 - \alpha^2)^{1/2}$ and $1/t = [d(1 - \beta^2)^{1/2} - a]/d\beta$.

Rearranging these two equations yields

$$\alpha = t(1 - \alpha^2)^{1/2} + b/c, \tag{2.16}$$

$$\beta = t[(1 - \beta^2)^{1/2} - a/d]. \tag{2.17}$$

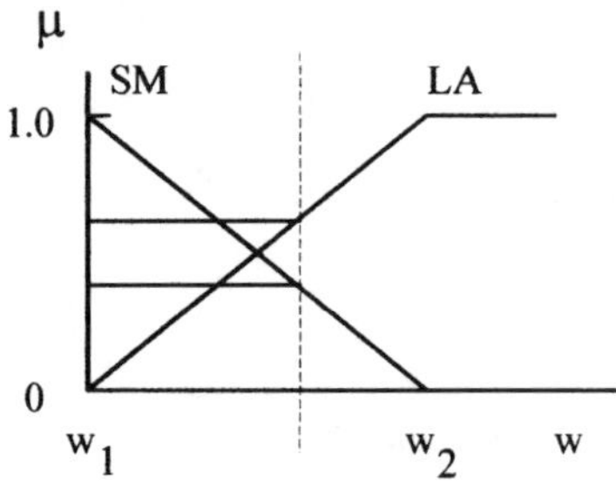

Figure 2.3. Partitioning of the w UD. SM = Small. LA = Large.

For given values of a, b, c, d and t, Equations (2.16) and (2.17) may be solved by numerical methods.

The transition curve is defined by $r(w)$. Both r and w are functions of $c(u)$ and $d(v)$, modified by membership value functions. In the FL treatment, the angle w represents the elements of the continuous array, $0 \leq w \leq \pi/2$. The UD is partitioned into two fuzzy sets as shown in Figure 2.3. The transition curve touches the circular arcs at $w = w_1$ and w_2, and the transition curve and circular arcs have common tangents at these points. The circular arcs are considered to be fuzzy sets on the w UD, as illustrated in Figure 2.3.

The corresponding membership functions are

$$\mu_{SM} = (w_2 - w)/(w_2 - w_1) \quad \text{and} \quad \mu_{LA} = (w - w_1)/(w_2 - w_1).$$

Referring to Figure 2.2, the rectangular co-ordinates of a point on the transition curve are obtained by the sum of the fractional parts of the co-ordinates (x, y) of the circular arcs, the fractions being controlled by the membership functions. Hence, within the transition zone,

$$x = \mu_{SM}\, h + \mu_{LA}\, g, \tag{2.18}$$

$$y = \mu_{SM}\, e + \mu_{LA}\, f, \tag{2.19}$$

where $e = d \sin v$, $g = c \cos u$, $f = c \sin u - b$ and $h = d \cos v - a$.

EXAMPLE 2.6. Assume that two coplanar arcs of 4 and 5 units radius, as shown in Figure 2.2, are to be blended by a smooth transition curve between $w = 10$ deg and $w = 30$ deg. The axes of the arcs are perpendicular and the arc centres are at $a = 1$ unit and $b = 3$ units. Find suitable polar co-ordinates for the transition curve.

Solution. From Equations (2.16) and (2.17)

$$\alpha = t(1 - \alpha^2)^{1/2} + 0.6$$

and

$$\beta = t[(1 - \beta^2)^{1/2} - 0.25].$$

Selecting successive values of $w = 10, 15, 20, 25$, and 30 deg and solving the above two equations numerically gives the following results for the angles u and v deg:

w **deg**	α	β	u**deg**	v **deg**
10	0.7217	0.1310	46.20	7.527
15	0.7709	0.1958	50.40	11.29
20	0.8126	0.2605	54.35	15.10
25	0.8463	0.3230	57.81	18.84

The rectangular co-ordinates of the two circular arcs may thus be found from formulae (2.18) and (2.19),

$$f = 5 \sin u - 3, \quad g = 5 \cos u,$$

$$e = 4 \sin v, \quad h = 4 \cos v - 1.$$

Thus,

w **deg**	f	g	e	h
10	0.6085	3.461	0.5240	2.966
15	0.8545	3.187	0.7832	2.923
20	1.063	2.914	1.042	2.862
25	1.232	2.664	1.282	2.786
30	1.373	2.424	1.550	2.687

From the given values of the angle w and Figure 2.3 the membership values are found and thus the Cartesian and polar co-ordinates of the transition curve,

w **deg**	μ_{SM}	μ_{LA}	$\mu_{\text{SM}}^{1.045}$	$\mu_{\text{LA}}^{1.045}$
10	1.0	0.0	1.0	0.0
15	0.75	0.25	0.7404	0.2349
20	0.50	0.50	0.4846	0.4846
25	0.25	0.25	0.2349	0.7404
30	0.0	0.0	0.0	1.0

w **deg**	$n = 1$		$n = 1.045$		
	x	y	x	y	r
10	2.974	0.524	2.966	0.524	3.912
15	2.989	0.801	2.913	0.7807	3.016
20	2.888	1.053	2.799	1.020	2.979
25	2.695	1.247	2.626	1.216	2.894
30	2.424	1.373	2.424	1.373	2.785

The value of r in the above table is for $n = 1.045$. The co-ordinates in this case provide a smoother transition curve than for those with $n = 1.0$. This is an example of hedging the membership values.

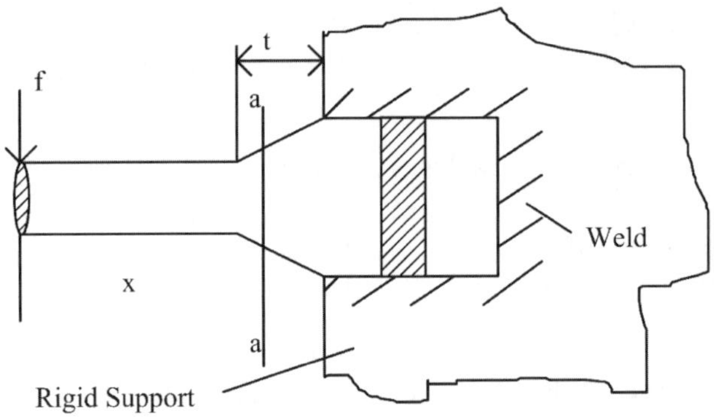

Figure 2.4. Loaded cantilever welded to a rigid support.

2.7. Fairing Solid Sections

It is often the case in engineering design that two sections of different geometry are to be blended together. Moreover, there may also be the requirement to satisfy strength, hydrodynamic or some other criterion. Such constraints must be considered in formulating the transition between the adjacent sections.

Consider for instance, the case of a cantilever carrying a transverse end load, f, as shown in Figure 2.4. The loaded end section is circular and the opposite rectangular section end is welded to a rigid support. The two ends are joined by a transition zone as shown.

The load (f) produces a bending moment (m) at any cross-section distance, x, from the free end,

$$m = fx. \tag{2.20}$$

If the transition zone is to be of uniform resistance to bending, then according to basic solid mechanics theory, the section modulus should increase uniformly in the transition zone. If the component is critical in a refined design, finite element or other numerical method could be applied to the problem of determining the transition shape. The first approximation shape can also be used as a basis for the finite element approach.

At some section a–a within the transition distant x from the free end, the geometry of the section will comprise both circular and rectangular features, as illustrated in Figure 2.5.

From standard solid mechanics theory, the second moment of area (I) of the free end circular section is

$$I_{\text{LD}} = \pi r^4/4 \tag{2.21}$$

(I is the conventional symbol for the second moment of area).

Also, the opposite end rectangular section second moment of area (I_{HI}) is

$$I_{\text{HI}} = bd^3/12. \tag{2.22}$$

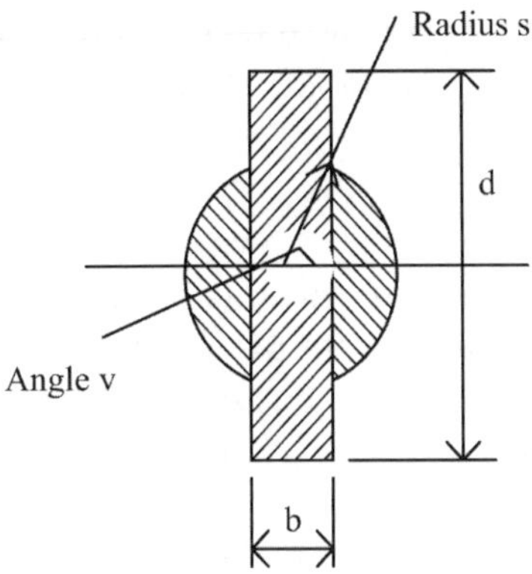

Figure 2.5. Typical transition zone section (d, v and s are all functions of x, b is constant).

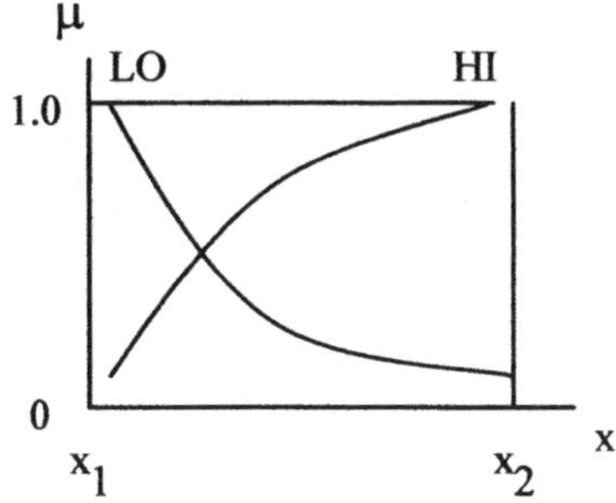

Figure 2.6. Membership function distribution in the transition zone.

For the sector of a circle, the second moment of area about a transverse axis through its centre is

$$I_S = s^4/6\{3v/2 - \sin 2v + (\sin 4v)/8\}, \tag{2.23}$$

where s is the radius of the sector from the centre of the section.

Now the second moments of area I_{LO} and I_{HI} may be regarded as interpenetrating fuzzy sets on the x UD, as illustrated in Figure 2.6.

To allow flexibility in the choice of section shapes, general forms of the membership functions may be assumed

$$\mu_{\mathrm{HI}} = k(x - x_1)^n \quad \text{and} \quad \mu_{\mathrm{LO}} = 1 - \mu_{\mathrm{HI}} \quad (n < 1).$$

The second moment of area of the whole transition zone section may then be portrayed as

$$I(x) = (\mu_{\mathrm{LO}} I_{\mathrm{LO}} + \mu_{\mathrm{HI}} I_{\mathrm{HI}}). \tag{2.24}$$

Now $\mu_{\mathrm{LO}} I_{\mathrm{LO}}$ represents the contribution of the circular component to the total second moment of area, $i(x)$, i.e., the contribution of two sectors of a circle, as in Figure 2.5, therefore,

$$2I_S = \mu_{\mathrm{LO}} I_{\mathrm{LO}}. \tag{2.25}$$

The value of the radius of the sector (s) at each section may be found by iteration of Equation (2.23), for a given value of I_S from Equation (2.25). The following example outlines the method.

EXAMPLE 2.6. A cantilever is transversely loaded, as shown in Figure 2.4. It has a circular section at the free end of 20 mm radius which is to be faired into a rectangular section of 10 mm breadth and 70 mm depth over a distance (x) of 45 mm. The membership functions for the transition may be assumed to be of the form shown in Figure 2.6, in which $k = 0.03864$ and $n = 0.8547$.

Find suitable forms of the transition zone sections at 5 mm intervals. Also check whether the transition zone dimensions provide a satisfactory bending strength.

Solution. From the given data, the membership functions are

$$\mu_{\mathrm{HI}} = 0.03864(x - x_1)^{0.8547} \quad \text{and} \quad \mu_{\mathrm{LO}} = 1 - \mu_{\mathrm{HI}}.$$

Also from the given data and Equations (2.21) and (2.22), the second moments of area of the end sections are

$$I_{\mathrm{LO}} = 1.2757 * 10^5 \text{ mm}^4 \quad \text{and} \quad I_{\mathrm{HI}} = 2.858 * 10^5 \text{ mm}^4.$$

The value of the component second moments of area $\mu_{\mathrm{LO}}I_{\mathrm{LO}}$ and $\mu_{\mathrm{HI}}I_{\mathrm{HI}}$ at each interval ($x - x_1$) may therefore be calculated. Assuming that a constant breadth $b = 10$ mm of the rectangular component (Figure 2.5), then the depth for each section may be found from Equation (2.22). The results of these calculations are tabulated below:

$(x - x_1)$ mm	0	10	20	30	40	45
μ_{LO}	1.0000	0.7235	0.5000	0.2928	0.0957	0.0000
μ_{HI}	0.0000	0.2765	0.5000	0.7072	0.9043	1.0000
$\mu_{\mathrm{LO}}I_{\mathrm{LO}} * 10^{-5}$ mm^4	1.257	0.9094	0.6285	0.3680	0.1203	0.0000
$\mu_{\mathrm{HI}}I_{\mathrm{HI}} * 10^{-5}$ mm^4	0.0000	0.7902	1.429	2.021	2.584	2.858
d mm	40.00	45.58	55.53	62.33	67.60	70.00
b mm	10.00		constant			

The value of the sector radius, s, (Figure 2.5), at each section is found by applying Equations (2.23) and (2.25), starting at the circular end section ($x - x_1) = 1$, where $s = 20$ mm. The iterations proceed as follows.

At ($x - x_1) = 10$ mm. Assume an initial radius value, $s = 20$ mm, also given that $b/2 = 5.0$ mm, then the angle $v = \cos^{-1}(5/20) = 75.52$ deg or 1.318 rad.

Therefore,

$$\sin 2v = 0.4842 \quad \text{and} \quad \sin 4v = -0.8473.$$

From Equation (2.23),

$$6I_S/s^4 = 3 * 1.318/2 - 0.4842 - 0.8473/8 = 1.387.$$

But from the above tabulated data I_S $(= \mu_{\text{LO}} I_{\text{LO}}) = 0.9094 * 10^5/2 = 0.4547 *$ 10^5 mm^4. Hence,

$$s^4 = 6 * 0.4547 * 105/1.387$$

or $s = 21.06$ mm.

Now performing a second iteration at $(x - x_1) = 10$ mm with the new value of $s = 21.06$ mm. The angle $v = \cos{-1}(5/21.06) = 76.27$ deg $= 1.331$ rad.
Therefore,

$$\sin 2v = 0.4611 \quad \text{and} \quad \sin 4v = -0.8184,$$

$$6I_S/s^4 = 3 * 1.331 - 0.4611 - 0.8184/8 = 1.433$$

Hence,

$$s^4 = 6 * 0.9094 * 105/2 * 1.433,$$

$$s = 20.89 \text{ mm}.$$

The third iteration gives $s = 20.91$ mm. The difference is then sufficiently small for most design purposes for this value to be accepted. Corresponding values of s for each $(x - x_1)$ interval may be found, these values are tabulated below. The total second moment of area and section modulus are also included.

$(x - x_1)$ mm	0	10	20	30	40	45
s mm	20	20.91	19.30	17.24	13.72	5.0
$I * 10^5$ mm	1.257	1.700	2.058	2.389	2.704	2.858
$z * 10^5$ mm	6.285	7.458	7.410	7.666	8.000	8.160

If the end sections have adequate strength, then the section modulus (z) shows that the other sections have a margin of reserve strength.

2.8. Remarks

In the present context, Euclidean geometry is unusual, it provides a corpus of knowledge comprising perfect conceptual objects and exact relationships within a framework of classical (Aristotelian) logic. There is no empirical content to this and no scope for the application of fuzzy logic, unlike the semi-theoretical and sometimes entirely empirical correlations of much of engineering science. Matching the imperfections of reality with Euclidean perfection is where fuzzy logic provides a unique bridge, because it can encompass uncertainty and vagueness in identity and also includes classical logic. It maintains the general shape of the Euclidean relations, whilst operating at the real world level with fuzzy numbers.

A metric for the distance apart of two rough surfaces would typically be chosen as the normal distance between the centre arithmetic mean line (Standard BS

1134:Part 1: 1988) of each surface. The metric is then given as a precise number. In the FL representation the metric would be a fuzzy number, which contains more information than a single number. The question of the distance apart of rough surfaces or boundaries then depends upon the membership level in the fuzzy set.

Abstract articulated linkages, such as the ubiquitous four-bar chain assembly, have theoretically precise loci at each point on the mechanism. This is achieved by geometric (positive) constraint such as in a journal bearing, or force closure such as in a cam follower. If there is force closure, then the possibility exists that the force may be deficient and that the closure may not be exact. There is also the possibility in this instance and in geometric constraint that design factors, wear and mal-assembly, for example, may affect the mechanism's performance and produce fuzzy loci. If the mechanical closure is not fully effective, then the relationships between the parts of the system become imprecise and even empirical, which results in a locus of motion generally represented by a subnormal fuzzy set.

Chapter 3
Material Selection

The selection of the most suitable material for a particular application is a crucial function in the design of an artefact. The performance of materials is important, not only under service conditions, but also in processing and in its economic and environmental effects. The characteristics of a material fundamentally influence the design geometry of an artefact and its compatibility with the service demands.

Obviously, a wide range of physical properties needs to be addressed in a comprehensive review of materials, which should also include economic factors. Indeed, so wide is the range that only a limited selection can be presented here, but principles of materials assessment based upon FL methods are applied to several prototype examples to provide patterns for further applications.

There are several complicating factors to be kept in mind. Some physical properties of engineering interest, such as Young's Modulus may show significant directional properties. This anisotropy is apparent, for example, in reinforced fibre materials and in wood. Cold working of metals also induces anisotropy. There are other factors too, such as scale effects, whereby a material may show different strength properties in the bulk to that displayed in relatively small test pieces. Also, surface stresses are two-dimensional, whereas it is the strains that are two-dimensional on planes of symmetry in the core. This means that the stress fields in laboratory test pieces should resemble those found in real artefacts in establishing failure criteria. With the above caveats in mind it is possible to set forth the principles of materials selection based upon simple material properties. In practice, it may be necessary to express the performance in terms of fuzzy numbers.

The selection of a specific material for a particular application is on the basis of its rating compared with other possible contenders. This leads to a search for an index which will most effectively summarise the compatibility of the material with the envisaged service demands. The relevant material performance is expressed in terms of several material properties, such as; stiffness, strength, weight, ductility and cost. To summarise the material properties the various factors are often evaluated and weighted according to their criticality and aggregated into a single index to provide a basis for comparison.

Table 3.1. A sample classification of some common engineering properties.

Mechanical	Electrical	Thermal and Radiation	Surface	Miscellaneous
Young's Modulus	Dielectric Strength	Specific Heat	Texture	Cost
Yield Stress	Resistivity	Conductivity	Corrosivity	Reliability
Toughness	Permativity	Expansivity	Wear Resist.	Durability
Ductility		Diffusivity		Recycleability
Density		Transmissivity		Manufacturability
Fatigue		Reflectivity		
Resistance		Emissivity		
Compressibility				
Hardness				
Creep Resistance				

3.1. Performance Parameters

The initial step in describing material performance is to classify relevant engineering properties in the types of phenomena. This would produce an extensive compilation of bulk and surface, physical and chemical properties. Economic factors should also be include in a fully comprehensive list. However, economic factors change over time and also from one country to another. Within the restricted range of engineering solids there are metals, polymers, elastomers, ceramics and glasses, each with numerous types and each of which passes through the liquid or powder state during processing prior to being used in the solid state. This still leaves a substantial volume of properties to consider. A sample classification list of engineering properties is shown in Table 3.1.

The above physical properties refer to solids. Liquids such as paints, lubricants and coatings would have other properties.

If there is only one material criterion to satisfy, say stiffness, then the selection of the material with the highest Young's modulus would be the obvious choice. However, such a simple case is unlikely in practice, there are invariably other factors such as brittleness and cost which would also influence the final choice.

A useful comprehensive approach to the identification of the most suitable material type to satisfy given criteria is available in the form of Ashby charts for properties of engineering solids. These charts group materials into broad categories for various combinations of properties. An example is shown in Figure 3.1, which displays Young's modulus against material density. On this chart, if there is a constraint of a minimum stiffness then a horizontal line drawn across the chart divides those materials that are able to satisfy the constraint from those that are not.

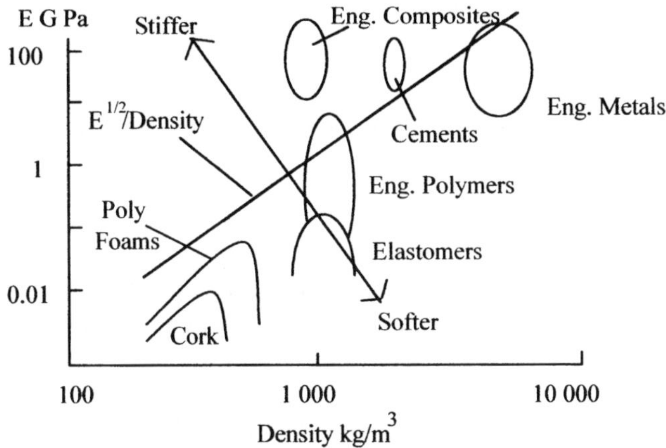

Figure 3.1. An Ashby chart of Young's Modulus against density for engineering solids.

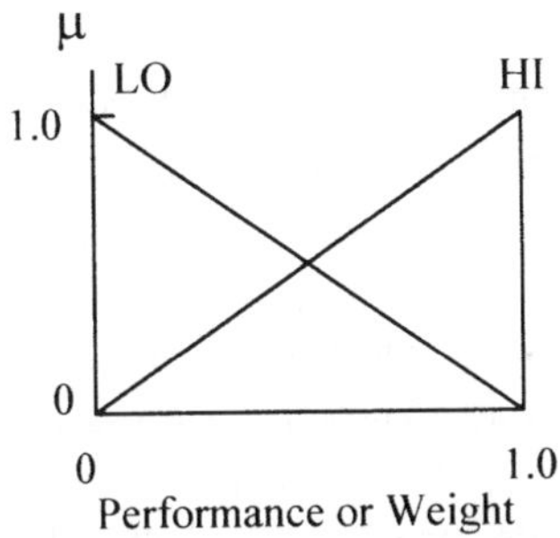

Figure 3.2. Partitioning of the performance and weight. LO = Low, HI = High.

In the aerospace, automobile and other industries there is also a requirement to minimise weight. In this case a criterion including density is required.

A useful feature of the chart shown in Figure 3.1 is that on the double log plot a line drawn with a slope of 1/2 identifies materials with equal stiffness per unit mass in simple beam bending. Those lying above the line have greater stiffness per unit mass, whilst those below the line have smaller stiffness per unit mass. This may be proved as follows. Consider a simply supported uniform cylindrical beam, length l and diameter d, under the action of a central transverse load w, as shown in Figure 3.2.

Simple beam theory gives an expression for the central deflection, (δ), under the load w,

$$\delta = wl^3/48EI, \tag{3.1}$$

where I is the second moment of area of the circular section ($I = \delta d^4/64$). With the usual notation the beam stiffness is therefore,

$$w/\delta = 3\pi d^4 E/4l^3. \tag{3.2}$$

The mass of the beam (m) is given by

$$m = \rho \pi d^2 / 4, \tag{3.3}$$

where ρ is the material density. Eliminating d between Equations (3.2) and (3.3) yields

$$(w/\delta)^{1/2}/m = 8(3/l^5)^{1/2}(E^{1/2}/\rho). \tag{3.4}$$

If the length is constant, then the maximum stiffness/unit mass is obtained when $(E^{1/2}/\rho)$ is a maximum for simple beam bending. Which verifies the above assertion.

By a similar analysis for the same physical case of a centrally loaded simply supported beam, but of ductile material with a yield stress of σ_y, it may be shown that

$$(wl/4)^{2/3}/m = 4(3/2\pi)^{2/3}(1/l)(\sigma_y^{2/3}/\rho). \tag{3.5}$$

This means that if the length of the beam is constant, then for various materials, the maximum sustainable bending moment per unit mass is obtained when $\sigma_y^{2/3}/\rho$ is maximised. Material selection may be conducted on a chart of $\log \sigma_y$ against $\log \rho$ with lines of 2/3 slope.

It follows that $E^{1/2}/\rho$ and $\sigma_y^{2/3}/\rho$ are two possible criteria on which to base material selection, depending upon design specifications. There are similar published criteria for a number of other important cases. To minimise mass in structural applications for a given level of stiffness, ductile strength or cleavage (brittle) strength, the general form of the criterion is cE^n/ρ, σ_y^n/ρ or k_{IC}^n/ρ respectively, where the index n depends upon the material geometry and the type of loading. c is a numerical factor. k_{IC} is the critical stress intensity factor for mode 1 {opening mode} failure. Values of c and n for a number of cases are shown in Table 3.2. It is assumed in this type of analysis that to a sufficiently good approximation, $G = 3E/8$ and $\nu = 1/3$.

For a spring of minimum volume or mass, the maximising criterion is σ_y^2/E. Obviously, for a load limited displacement free design σ_y or k_{IC} should be maximised, but for a displacement limited but load free design the criterion is σ_y/E or k_{IC}/E for ductile or brittle materials respectively. A ductile material is one that deforms plastically before fracture, and for this condition k_{IC}/σ_y is the appropriate criterion.

3.2. Comparison of Material Criteria

There are several steps to the selection of materials for particular applications. The first of these is the choice of the type of material most suited to the service environmental conditions, such as operating temperature, corrosion attack, wear, shock loading, and material and manufacturing costs. At the same time innovative

Table 3.2. A summary of c and n values for minimum weight.

Case	Stiffness (n, c)	Strength Ductile/Brittle
Tie bar	1, 1	1
Torsion bar or tube	1, 3/8	1
Beam bending	1, 1/2	1/2
Column buckling	1, 1/2	–
Plate bending	1, 1/3	1/3
Plate buckling	1, 1/3	–
Internally pressurised cylinder	1, 1	1
Rotating cylinder	1, 1	1
Internally pressurised sphere	3/2, 1	1

design solutions should be considered with new material combinations and finishes. The preliminary consideration will eliminate large classes of materials and there remains those worthy of consideration in more detail.

The next step is the consideration of material criteria. The comparison of material performance using criteria such as those listed in Table 3.2 provides a broad design guide. It will be noted however, that the criteria postulate the same type of artefact geometry for each type of material considered, whereas the material type can influence the geometry for the same service conditions. Clearly, the optimising criteria are most useful where the geometry (which may also depend upon the type of manufacturing process) is not much affected by the material type. In practice, the number of criteria should be limited to six or less to mitigate the effect of averaging out of performance.

A study of the performance criteria, such as those listed in Table 3.2, within an Ashby chart (see Further Reading) can reveal some interesting information. For example, woods can offer greater stiffness/unit mass than metals and can also be equally effective as leaf springs. Of course, there are other factors that may detract from the use of wood. Some material properties may be evaluated more precisely than others, for example, Young's modulus compared with the yield stress. The latter is often a matter of personal judgement. Another factor is that different grades of the same type of material or the same material with different manufacturing processes, can affect the properties. For example, shot peening a metal surface can markedly change its fatigue resistance. This means that in some cases there can be a lack of precision in the property values, which are then most naturally represented by fuzzy numbers to avoid over-representing the state of knowledge.

With the above caveats, it is possible to make broad comparisons of materials. An example is given below.

EXAMPLE 3.1. Compare the beam bending stiffness, the plate bending or buckling stiffness and beam bending ductile strength per unit mass of acrylic, aluminium alloy and mild steel at room temperature.

Solution. Typical mechanical properties of acrylic are:

$$E \text{ (GPa)} = 0//2.6 + 0.5//2.7 + 1.0 + //2.95 + 0.5//3.2 + 0//3.2,$$

$$y \text{ (MPa)} = 0//45 + 0.5//50 + 1.0//65 + 0.5//80 + 0//85,$$

$$\rho \text{ (kg/m}^3) = 120.$$

(This fuzzy set has one element.)

(i) Beam bending stiffness/unit mass. Using the above data and the information in Table 3.2 yields

$$E^{1/2}/\rho = 0//0.0134 + 0.5//0.0137 + 1.0//0.0143 +$$
$$+ 0.5//0.0149 + 0//0.0152.$$

(ii) Plate bending or buckling stiffness/unit mass

$$E^{1/3}/\rho = 0//0.0115 + 0.5//0.0116 + 1.0//0.0120 +$$
$$+ 0.5//0.0123 + 0//0.0124.$$

(iii) Beam bending ductile strength/unit mass

$$\sigma_y^{2/3}/\rho = 0//0.1054 + 0.5//0.1131 + 1.0//0.1347 +$$
$$+ 0.5//0.1547 + 0//0.1611.$$

Typical mechanical properties of aluminium alloy are:

$$E \text{ (GPa)} = 0//69 + 0.5//70 + 1.0//71 + 0.5//72 + 0//73,$$

$$\sigma_y \text{ (MPa)} = 0//250 + 0.5//300 + 1.0//350 + 0.5//400 + 0//450,$$

$$\rho \text{ (kg/m}^3) = 0//2626 + 0.5//2667 + 1.0//2708 + 0.5//2749 + 0//2790.$$

In this case the optimising criteria are found by Cartesian products. The Cartesian product, $E^{1/2} \times 1/\rho$ yields the following arrays for aluminium alloy; (all SI units)

$$E^{1/2} \times 1/\rho = \begin{vmatrix} 0 & 0 & 0 & 0 & \mathbf{0} \\ 0 & 0.5 & 0.5 & \mathbf{0.5} & 0 \\ 0 & 0.5 & \mathbf{1.0} & 0.5 & 0 \\ 0 & \mathbf{0.5} & 0.5 & 0.5 & 0 \\ \mathbf{0} & 0 & 0 & 0 & 0 \end{vmatrix} // \begin{vmatrix} 3.16 & 3.11 & 3.06 & 3.02 & \mathbf{2.97} \\ 3.19 & 3.14 & 3.09 & \mathbf{3.04} & 3.00 \\ 3.21 & 3.16 & \mathbf{3.11} & 3.07 & 3.02 \\ 3.23 & \mathbf{3.18} & 3.14 & 3.09 & 3.04 \\ \mathbf{3.25} & 3.20 & 3.15 & 3.11 & 3.06 \end{vmatrix} * 10^{-3}.$$

The principal values in the above arrays are shown in bold type, they provide the following fuzzy number:

$$E^{1/2}/\rho = [0//2.97 + 0.5//3.04 + 1.0//3.11 +$$
$$+ 0.5//3.18 + 0//3.25] * 10^{-3},$$

where the 10^{-3} refers only to the physical values in the above expression.

Similarly, the principal values resulting from the Cartesian product, $E^{1/3} \times 1/\rho$ gives (SI)

$$E^{1/3}/\rho = [0//1.438 + 0.5//1.499 + 1.0//1.529 +$$
$$+ 0.5//1.560 + 0//1.591] * 10^{-3}.$$

Also, the principal values resulting from the Cartesian product $\sigma_y^{2/3} \times 1/\rho$ gives

$$\sigma_y^{2/3}/\rho = [0//1.423 + 0.5//1.631 + 1.0//1.834 +$$
$$+ 0.5//2.036 + 0//2.237] * 10^{-2}.$$

Finally, considering mild steel. The mechanical properties are more sharply defined as follows:

$$E = 207 \text{ GPa}; \quad \sigma_y = 280 \text{ MPa}; \quad \text{and} \quad \rho = 7850 \text{ kg/m}^3.$$

(These fuzzy sets have one element each.)

These values provide the following criteria (SI):

$$E^{1/2}/\rho = 1.832 * 10^{-3}, \quad E^{1/3}/\rho = 7.534 * 10^{-4}, \quad \sigma_y^{2/3}/\rho = 5.453 * 10^{-3}.$$

The three materials discussed above have widely disparate mechanical properties. At room temperature, the acrylic material is best on each criterion, whilst mild steel is worst. But as the temperature is increased the order would be reversed. Another factor is that polymers, unlike metals, often show significant time-dependency in their mechanical properties, which becomes more pronounced as the temperature rises.

A summary of the performance of the three materials is given below using the principal values ($\mu = 1$) from the fuzzy numbers.

Material	$E^{1/2}/\rho$	$E^{1/3}/\rho$	$\sigma_y^{2/3}/\rho$	σ_y^2/E
Acrylic	0.0143	0.0120	0.1347	1432
Al. Alloy	0.0031	0.0015	0.0183	1725
Mild St.	0.00183	0.00075	0.00545	379

The final column in the above list represents the maximum energy content of a spring per unit volume. The aluminium alloy appears to be the best on this criterion and again mild steel the worst.

3.3. Material Performance Grading

In practice it would be rare for material selection to be based upon just one criterion, more commonly it is a question of selecting the least number of criteria that would adequately represent the total ability of the material to satisfy all the service and economic requirements and also the wider interests of the manufacturers and consumers. The problem of selecting the most suitable material for an application involves a decision about the most favourable balance of the selected criteria. The onus of the decision can be removed from the designer, to a certain extent, by an agreed weighting of each of the criteria according to its relative importance, which means a management input. The onus may be more fully removed by a method of aggregating or combining the criteria and weighting to provide an overall material performance index (MPI). In effect, this moves the decision upstream in the information processing system.

One form of aggregation to find an MPI is by summation of the product of performance and weight for each criterion, thus,

$$MPI = \sum_i performance(i, j) * weight(j). \tag{3.6}$$

Aggregation by the product of products is an alternative to the above method,

$$MPI = \prod_i performance(i, j) * weight(j). \tag{3.7}$$

Forms more suited to FL methods are given later; see Equations (3.9) and (3.10).

Hopgood (see Further Reading) has opined that neither of these two methods of aggregation are satisfactory because in Equation (3.6), a poor performance by an important criterion (high weighting) may be balanced by the high performance of a relatively less important criterion. In the case of Equation (3.7) low product values, including low performance with high weight and high performance with low weight, tend to be masked, leaving the final selection to a good average performance. A material with both a low performance and low weighting would have some good performance values with good weighting.

The above deficiencies in the simple summation and product aggregation methods has prompted the development of another aggregation method by Hopgood. This is based upon the product of products, Equation (3.7), but with the empirical numerical adjustments of offset and scale shift terms, as shown in Equation (3.8),

$$MPI = \prod_i \{[weight(j)(performance(i, j) - offset)]scale\ shifting\ term\}. \tag{3.8}$$

Table 3.3. AIM scores for several perform-
ance/weight combinations.

Performance	Weight	AIM Score
0	0	0.46
0	1	0.46
1	0	0.01
1	1	1.00

This is called the AIM score method. By adjustment of the numerical terms negative values are avoided and so also is the undue influence of poor performance
with low weight combinations. The AIM score for a single criterion of a polymeric
material on the basis of Equation (3.8), in which the performance and weight are
both divided into two classes, high and low, is given in Table 3.3.

An unusual feature of the AIM score is its value for a low performance and
weight. Values of 0.45 and 0.46 are assigned to the offset and scale shift terms
respectively in Equation (3.8).

The AIM score may be displayed in the form of a three-dimensional graph with
axes: Weight, Performance and AIM score (0–1). Although the AIM score method
does appear to achieve some of the desired effects, it does so by the introduction of
empirical numbers that give some impression of precision, but are infact arbitrary.
An alternative is the fuzzy aggregation approach described below, which avoids
overstating the precision of the assumptions.

3.4. Fuzzy Aggregation

Adjustment of the empirical aggregation formula, Equation (3.8) by means of arbitrary factors presents the state of knowledge in an unrealistically precise form. The
need for this arises from a too early introduction of a deterministic description at a
stage when only broad trends are important. It will be shown here how a FL relational base can provide a palette of rules that enables conclusions to be drawn, and
ultimately decisions to be made without an early commitment to precisely defined
formulae. The framework is not prescriptive in detail, only general tendencies are
prescribed.

To construct the FL framework, the logic spaces (UDs) of performance and
weight are defined and also that of the criterion score. In the present work all three
types of UDs will be defined in the interval, 0–1.0. They are partitioned into two
triangular fuzzy sets each in the case of performance and weight and into four
triangular fuzzy sets in the case of the criteria scores. The partitioning for the three
cases is illustrated in Figures 3.2 and 3.3, respectively.

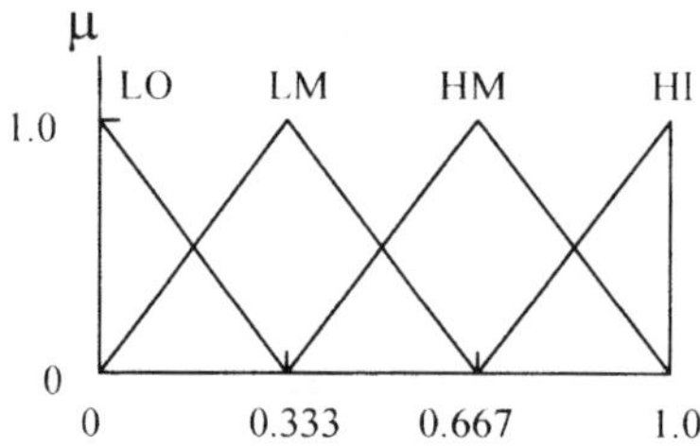

Figure 3.3. Partitioning of the criteria score. LO = Low, LM = Low Medium, HM = High Medium, HI = High.

Table 3.4. A relational rule base for material scores.

Performance	Weight	
	L	**H**
L	LM	LO
H	HM	HI

Where L = Low, H = High, and LO = Low, LM = Low medium, HM = High medium, HI = High.

The membership functions for the partitioning in Figure 3.2 are

$$0 \leq x \leq 1.0$$

$$\mu_{LO} \quad 1 - x$$
$$\mu_{HI} \quad x$$

where x denotes performance (p) or weight (w), as the case may be.

The membership functions for the partitioning in Figure 3.3 are

	$0 \leq s \leq 0.333$	$0 \leq s \leq 0.667$	$0.667 \leq s \leq 1.0$
μ_{LO}	$1 - s/0.333$		
μ_{LM}	$s/0.333$	$2 - s/0.333$	
μ_{HM}		$(s - 0.333)/0.333$	$3 - s/0.333$
μ_{HI}			$(s - 0.667)/0.333$

The relationship between the material performance criteria, the relevant weight and the material score is expressed by a relational rule base, a simple example of which is shown in Table 3.4.

The material score fuzzy sets in Table 3.4 correspond with the partitioning illustrated in Figure 3.3.

The following example applies the methodology which is described above:

EXAMPLE 3.2. A particular polymeric solid has four performance criteria relating to a certain application. These values and the corresponding weightings are tabulated below:

Criterion	(a)	(b)	(c)	(d)
Performance score	0.8	0.8	0.2	0.2
Weight	0.8	0.2	0.8	0.2

Assume the partitioning of the UDs shown in Figures 3.2 and 3.3 and the relationship array in Table 3.4. Find the material score for each criterion and also the overall MPI.

Solution. For criterion (a) the respective performance and weight values are: $p = 0.8$ and $w = 0.8$. From Figure 3.2 or the membership function expressions, the following values may be found:

$$\mu_{PL} = 0.2 \mu_{WL} = 0.2,$$

$$\mu_{PH} = 0.8 \mu_{WH} = 0.8,$$

where the subscripts p and w refer to the performance and weight respectively.

The form of the fuzzy logic proposition relating the performance and weight antecedents to the score conclusion is

$$\text{IF} \quad \text{P AND W} \quad \text{THEN} \quad \text{S}$$

This proposition postulates an intersection operation. When applied to the present problem, it has the form

IF P	AND W	THEN S	MINIMUM	CONSEQUENCE
HP	HW	HI	0.8, 0.8	0.8HI
HP	LW	HM	0.8, 0.2	0.2HM
LP	HW	LO	0.2, 0.8	0.2LO
LP	LW	LM	0.2, 0.2	0.2LM

The conclusion is given by the union of the four consequences, i.e.,

$$S = 0.2LO \cup 0.2LM \cup 0.2HM \cup 0.8HI.$$

This conclusion is conveniently displayed in Figure Ex 3.1.

It may be noted from Figure Ex 3.1 that the HI set is the strongest contributor to the conclusion. Although Figure Ex 3.1 illustrates the full (fuzzy) solution, it is often desired to find a single representative (non-fuzzy) number There are several methods of defuzzifying such a solution, though they all involve some loss of information. Clearly, a single precise number cannot completely represent a fuzzy set. Methods of defuzzification are discussed in Chapter 1 of this text. Using the normalised weighting of the set principal values (see Equation (1.6)), gives

$$s_a = (0.2 * 0.333 + 0.2 * 0.666 + 0.8 * 1.0)/(0.2 + 0.2 + 0.8) = 0.833.$$

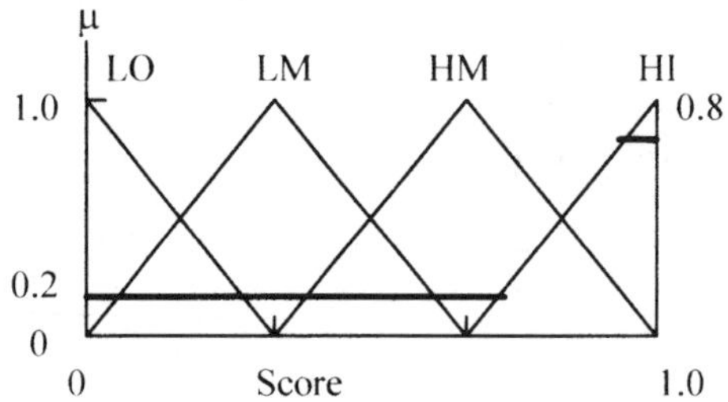

Figure Ex 3.1. Fuzzy score for $p = 0.8$ and $w = 0.8$.

Table Ex 3.1. Summary of polymer scores.

| | **Criterion** | | | |
Scores	**(a)**	**(b)**	**(c)**	**(d)**
Perf.*Wt.	0.64	0.16	0.16	0.04
DefS Wtd Set	0.833	0.799	0.500	0.500
Centroid[#]	0.663	0.569	0.356	0.407
AIM[##]	0.74	0.53	0.26	0.41

[#]By diagram scaling. [##]Offset 4.5, scale shift 20. Results reduced to a scale of 0–10.

The above process may be repeated for the remaining criteria, (b), (c) and (d). The results are displayed in Figure Ex 3.2 (b), (c) and (d).

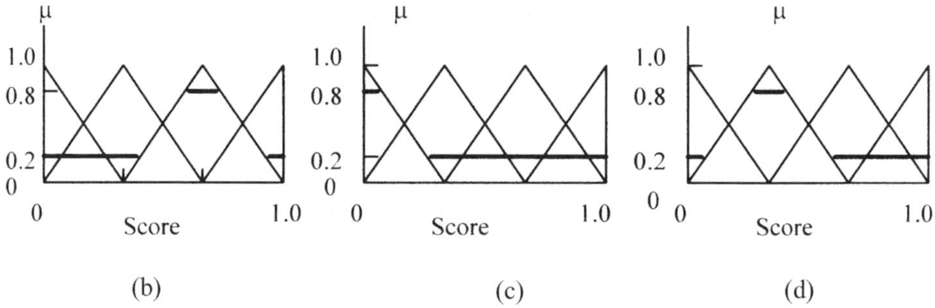

(b)   (c)   (d)

Figure Ex 3.2. Fuzzy scores for: (b) $p = 0.8$, $w = 0.2$. (c) $p = 0.2$, $w = 0.8$ and (d) $p = 0.2$, $w = 0.2$.

For these three scores the defuzzified values (normalised weighted set principal values) are $s_b = 0.799$, $s_c = 0.667$ and $s_d = 0.500$.

In the case where an overall MPI is needed for comparison of different materials, a rule for merging the scores for individual criteria is used. There are many possible arbitrary rules, typical of these are the arithmetic mean or the nth harmonic mean, Equations (3.9) and (3.10) respectively below, which are modified forms of Equations (3.6) and (3.7), and are applicable to both fuzzy and non-fuzzy aggregation,

$$\text{MPI}_{\text{av}} = \left(\sum \text{scores} \right) / n, \tag{3.9}$$

$$\text{MPI}_{\text{ha}} = \left(\prod \text{scores} \right)^{1/n}. \tag{3.10}$$

Applying Equations (3.9) and (3.10) to the four types of scores shown in Table Ex 3.1 yields the following MPIs, to two decimal places:

	Perf.*Wt.	Def S Wtd. Set	Def S Centroid	AIM
MPI_{av}	0.25	0.70	0.50	0.30
MPI_{ha}	0.16	0.69	0.48	0.22

It may be noted that the above treatment is free from any mathematical prescription as to how the performance and weight should be combined.

The AIM score is achieved with the insertion of an arbitrary offset and scale shift values and therefore has less meaning than the other three sets of values.

3.5. Material Benchmarking

New materials, new combinations of materials and new applications are continually evolving. In the search for continuous product improvement it is a familiar question whether an alternative material would be superior to that which is already in use in an application. This may be a crucial question if, for example, the artefact is a high volume production article and market share can be lost, or if there is the possibility of legal liability due to product failure. In either case, the value-at-risk is high and new material appraisal must be conducted in a systematic way.

The previously described MPI provides one way of making a value judgement by comparing the MPIs of possible alternative materials. However, this does require defuzzifying fuzzy conclusions which means the loss of some of their information content. Benchmarking provides an alternative way of formulating a value judgement as an aid to decision making. The method postulates a known material of proven performance and about which a value judgement is available for the benchmark material. This is specified by a team of several independent experts, based upon the specification of the service environment, loading pattern and economic circumstances. For example, let the management selected criteria be as follows:

(i) Finished cost of item (material and manufacturing) (CO).
(ii) Resistance to bending per unit mass (RE).

(iii) User satisfaction (US).
(iv) Corrosion resistance (CR).

Evaluation of the importance of each criterion for the existing material leads to a fuzzy discrete number in which the membership values express the consensus importance weighting of each of the criteria yielding set B below:

$$B = [\mu_{CO}//CO + \mu_{RE}//RE + \mu_{US}//US + \mu_{CR}//CR]. \tag{3.11}$$

Let the selected material provide the benchmark against which to compare other possible alternative materials. A panel of expert opinion can provide a value judgement of the total rating of the material against the criteria (i)–(iv) listed above. The polling of opinion will also provide a discrete fuzzy number, V, where,

$$V = [\mu_{IN}//PR + \mu_{FA}//FA + \mu_{GD}//GD + \mu_{SP}//SP], \tag{3.12}$$

where PR = poor, FA = fair, GD = good and SP = superior.

The question then is how to compare a new material with a certain criterion set with the benchmark set, V, and provide a comparative value judgement within the set, Equation (3.12). This is accomplished by creating an array which is a combination of the information contained in Equations (3.11) and (3.12), this is called the discriminator array (D). D is found from the Cartesian product of the fuzzy numbers B and V, thus,

$$D = V \times B = \begin{array}{r|cccc} & \textbf{PR} & \textbf{FA} & \textbf{GD} & \textbf{SP} \\ \textbf{LO} & \mu_{11} & \mu_{12} & \cdot & \cdot \\ \textbf{RE} & \mu_{21} & \cdot & \cdot & \cdot \\ \textbf{US} & \cdot & \cdot & \cdot & \mu_{34} \\ \textbf{CR} & \cdot & \cdot & \mu_{43} & \mu_{44} \end{array}$$

where $\mu_{11} = \min(\mu_{IN}, \mu_{LO})$, $\mu_{12} = \min(\mu_{FA}, \mu_{LO})$ and $\mu_{21} = (\mu_{IN}, \mu_{RE})$, etc.

Let a new material provide another set of criteria, B' corresponding with Equation (3.11). The value judgement of the new material, V', is obtained by the composition of B' with the discriminator array, D, thus let,

$$B' = [\mu'_{LO}//LO + \mu'_{RE}//RE + \mu'_{US}//US + \mu'_{CR}//CR]. \tag{3.13}$$

Now,

$$V' = B' \circ R = \text{maxmin}(\{\mu'_{LO}, \mu_{11}\}, \{\mu'_{RE}, \mu_{21}\}, \{\mu'_{US}, \mu_{31}\}, \ldots). \tag{3.14}$$

The result of the composition is another value judgement fuzzy set,

$$V' = [\mu_{IN}//IN + \mu'_{FA}//FA + \mu'_{GD}//GD + \mu'_{SP}//SP]. \tag{3.15}$$

This value judgement can be compared with that of the benchmark material, Equation (3.12) to decide whether or not it represents an improvement.

EXAMPLE 3.3. Development has been undertaken on an innovative plastic fuel tank for motor cars. Two versions have achieved prototype assessment status, they differ in the number of skin layers and their thickness. It is considered that the plastic tanks will be able to compete with the normal sheet steel tanks on price and may offer some advantages in mechanical properties and corrosion resistance. As part of the assessment process, benchmarking against the equivalent steel tank is required. For the materials section of the benchmarking exercise, four criteria are selected, namely,

(i)   Material cost (CO).
ii)   Bending strength/unit mass of the material (RE).
(iii) Impact resistance (SH).
(iv) Corrosion resistance.

The benchmark steel tank material has the following criteria weightings,

$$B = [0.56//CO + 0.3//RE + 0.55//SH + 0.32//CR].$$

The value judgement of the steel tank has the following description,

$$V = [0.2//IN + 0.4//FA + 0.55//GD + 0.3//SP].$$

The two plastic tank materials have the following material weightings,

$$B' = [0.38//CO + 0.63//RE + 0.48//SH + 0.48//CR],$$

$$B'' = [0.32//CO + 0.68//RE + 0.55//SH + 0.48//CR].$$

Compare the two plastic tanks with the steel tank on the basis of the materials.

Solution. From the material weightings and the value judgement set for the plastic tank, the discriminator array may be found:

		PR	FA	GD	SP
	CO	0.2	0.4	0.55	0.3
$D = V \times B =$	RE	0.2	0.3	0.33	0.3
	SH	0.2	0.4	0.55	0.3
	CR	0.2	0.32	0.32	0.3

To find the value judgement of the first plastic tank (V'), the maxmin composition is found, this yields

$$V' = B' \circ D = 0.2//PR + 0.4//FA + 0.48//GD + 0.3//SP.$$

Similarly, for the second plastic tank,

$$V'' = B'' \circ D = 0.2//PR + 0.4//FA + 0.55//GD + 0.3//SP.$$

The two plastic tanks have very similar value judgements, the second being slightly better that the first and as good as the steel tank. The choice of plastic or steel would have to be made on additional criteria, such as supply chain reliability.

3.6. Remarks

With the steadily widening variety of materials and their combinations available, the Ashby scheme of performance criteria formulation and the associated charts provides an invaluable broad classification method for most types of engineering solids. There is no equivalent at the present time for liquids and gases. The method could also be extended into non-engineering areas such as the medical and catering fields, each with their own specialist performance criteria.

In considering the Ashby charts, it will be recalled that material properties, and therefore performance criteria, are also affected by the material's service environment, such as temperature. This can have significant effects on some materials, especially plastics, and can considerably alter the landscape of the charts. Also, some materials, again particularly plastics, have time-dependent properties. Metals also exhibit creep at high temperatures.

Some material properties, such as Young's modulus for metals, may be quite precisely defined at a given temperature, but other properties, such as the yield point of ductile materials can only be approximately evaluated. This lack of precision can be naturally accommodated within a FL framework. This advantage clearly emerges in, for example, the combining of performance criteria with weighting. Various methods are in use at present, they all suffer from the disadvantage of being cast in deterministic mathematical form, which expresses a precision that is not supported by the state of knowledge. The FL method described does not portray exact relations and is therefore far more universal in its application.

Chapter 4
Hydrodynamic Lubrication

Intensive study of the physics and chemistry of lubrication, driven by the global demand for increasing mechanisation of human activities, has resulted in significant improvements in the efficiency, reliability and avialability of machines. This has partly resulted from improvements in lubrication technology. The improvements have had an influence on the conservation of the world's resources and quality of the environment through reduced energy wastage and rates of machine deterioration.

The modern study of lubrication is usually considered to have begun with the experimental work on loaded journal bearings by Beauchamp Towers in the latter part of the 19th century. Also with the slightly earlier study of friction in an unloaded bearing, resulting in the well-known formula by Petroff. Osbourne Reynolds applied laminar flow viscous flow theory to the observations of Beauchamp Towers and obtained the so-called lubrication approximation equations (which neglect fluid inertia effects). Reynolds drew some important basic conclusions from his studies:

(i)   Friction effects are due to fluid friction alone.
(ii) The load bearing capacity depends upon the lubricant viscosity.
(iii) The bearing journal is supported only by the lubricant film.

Reynolds analysis still forms the basis of most hydrodynamic lubrication theory for steadily loaded journal bearings. The Petroff formula only applies to the case where the journal is concentric with the bearing shell and is therefore unloaded. But it was proved by Reynolds that the journal must be eccentric relative to the shell for the practical case of a loaded bearing. Early in the 20th century, Sommerfeld derived a formula for the eccentricity of a loaded full film journal bearing in terms of the system operating parameters. The parameters were collected together in the form of a dimensionless group, now known as the Sommerfeld number. Sommerfeld's theory rests upon the same assumptions as those of Reynolds'.

All these early studies are widely reported in the engineering literature. They remain the basic lubrication theory for steadily loaded full film journal bearings and also for slider bearings operating at constant speed. Dynamic loads or fluctuating speeds requires extension to the basic theory, so too does the cavitating lubricant film. During start-up and shut-down of a machine, the solid bearing surfaces come into close contact, then lubrication depends more upon the properties of adsorbed molecular surface layers, derived from surface active molecules that are dispersed in the lubricant to reduce solid friction and wear of the surfaces. Under more severe conditions of load and contact, shearing of the surface asperities occurs, producing wear, hence the frequent use of a low yield stress material, such as white metal, on the one surface to provide sacrificial wear. Thus, full-film lubrication merges into the mixed and the so-called boundary lubrication regimes, before solid contact occurs.

Unlike the study of geometry, there is an empirical content to the relations found in lubrication studies, and substantial support is required from experimental work. The Engineering Sciences Data unit (ESDU), London, has issued a series of publication to summarise current engineering practice and to provide guide lines for the designer on lubrication technology. These publications provide formulae and charts spanning a wide range of operating parameters. The treatment is deterministic throughout, as in publications from other sources. Salient features of the published work are summarised below to form a basic study of trends in correlations.

4.1. Basic Concepts and Correlations

Lubrication theory usually commences with the case of two-dimensional flow in a plane slider bearing, which is extended to the journal bearing with a thin lubricant film. The journal bearing film is also treated as two-dimensional flow.

In the simple case of a steadily loaded journal bearing, the line of action of the load is perpendicular to the line joining the journal and bearing shell centres, as illustrated in Figure 4.1(a). This case is discussed at length in standard engineering texts. The eccentricity (e) is dependent upon the load (w) and it is zero for no load. The more general case is that of Figure 4.1(b), in which the line of action of the load is not perpendicular to the eccentricity direction. In this case there is cavitation within the fluid film, i.e., there is no longer full film lubrication. The extent of the cavitation is dependent upon the lubricant supply press and the pattern of the supply grooves within the bearing for a given load and bearing rotational speed.

For full film lubrication, Figure 4.1(a), equilibrium of moments gives

$$we + m_r = m, \tag{4.1}$$

where m is the shaft torque and m_r is the bearing reaction torque.

This provides an indirect means of measuring the eccentricity and therefore the minimum film thickness. Equation (4.1) does not apply to a cavitating film. In this case there is no simple direct or indirect means of estimating the eccentricity. Normally, to utilise the bearing capacity efficiently the dimensionless eccentricity ($2e/c$ where c is the diametrical clearance) would be designed to operate with $2e/c$ between 0.6 to 0.8. Too high a value promotes possible contact between surface

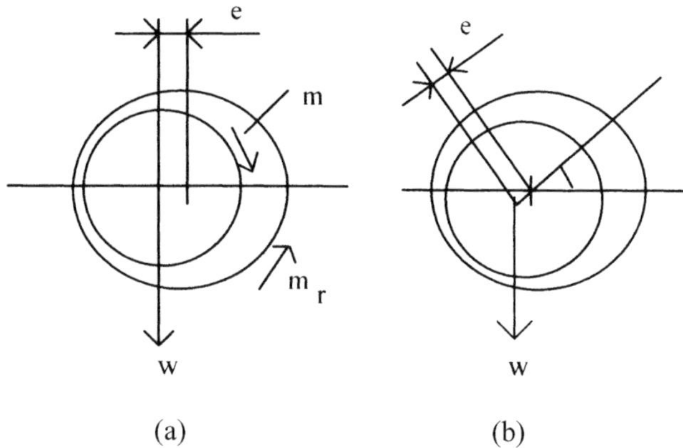

Figure 4.1. Journal bearing. (a) Full film. (b) Cavitating film.

asperities and therefore wear and energy waste in friction, whilst too low a value means that the bearing capacity is under-utilised.

Under real service conditions the texture of the two bearing surfaces may be of the order of 1.5 microns for a fine surface finish, but in practice may be significantly larger. There are also dimensional tolerances of roundness, straightness and parallelism as well as deflections due to load and thermal effects, and in addition wear. All of these factors can add up to significant differences in practice to the theoretical values and those used in research investigations.

The salient design features of a journal bearing performance are:

(i)   The virtual coefficient of friction.
(ii) The frictional power loss.
(iii) The minimum film thickness.

These are often expressed in terms of dimensionless parameters, the most common of these are:

(i)   Sommerfeld number, $s^+ = (d/c)^2 \eta n / p$, where $p = w/dl$, called the specific bearing pressure due to the load w. This parameter expresses the combined operating conditions. The reciprocal of the Sommerfeld number is called the dimensionless load.
(ii) Virtual friction coefficient, $\phi^+ = df^+/c$, where $f^+ = f/w$ and f is the peripheral friction force on the shaft.
(iii) Frictional power loss, $l^+ = \pi(df^+/c)/S^+$.
(iv) Minimum film thickness, $h^+ = 2h_0/c$ or $\varepsilon = 1 - h^+$, where h_0 is the minimum film thickness.

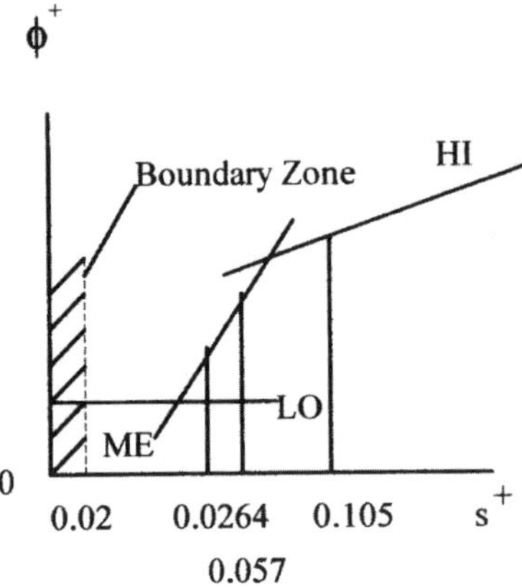

Figure 4.2. The virtual friction coefficient (ϕ^+) as a function of the Sommerfeld number (s^+).

Turbulent flow may occur in the fluid film under extreme conditions, the onset is governed by the fluid Reynolds number and the c/d ratio. Whirling of the shaft may also occur, depending upon the effective shaft mass, the bearing speed, the clearance c and the applied load w. These factors are not treated in this work.

The published data is often summarised in the form of charts and is condensed by the use of the dimensionless variables listed above. The data trends are outlined below.

4.1.1. VIRTUAL FRICTION COEFFICIENT

The boundary zone at the extreme left-hand side of Figure 4.2 is beyond the point where film lubrication conditions exist. To the right-hand side of this zone the virtual friction coefficient rises with increasing Sommerfeld number at and increasing, then a decreasing rate. The limiting theoretical case for high Sommerfeld numbers (load tending to zero), is

$$\phi^+ = 2\pi 2 s^+. \tag{4.2}$$

4.1.2. FRICTIONAL POWER LOSS

The frictional power loss is the source of lubricant heating which must be controlled within acceptable limits, by ensuring that it is effectively cooled. It is therefore a prime design factor. The dimensionless power loss (l^+) may be defined in terms of the virtual friction coefficient and the Sommerfeld number,

$$l^+ = \pi \phi^+ / s^+. \tag{4.3}$$

It may be noted that $\phi^+ = \phi^+(s^+)$, thus $l^+ = l^+(s^+)$. The general trend of l^+ is illustrated in Figure 4.3.

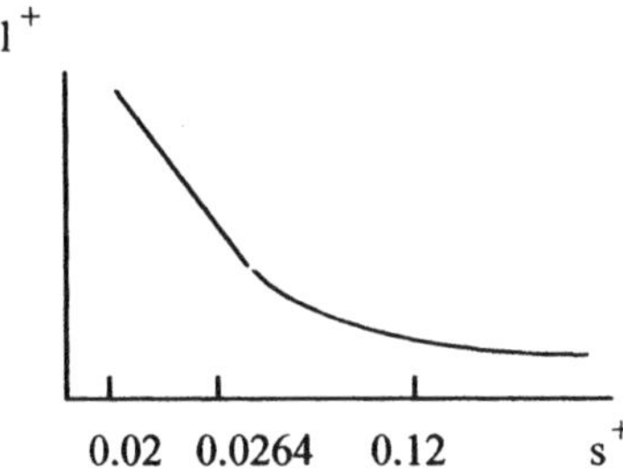

Figure 4.3. The trend of the $l^+(s^+)$ correlation.

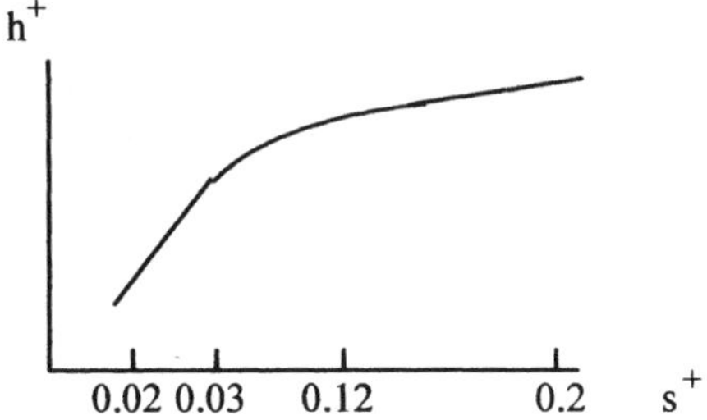

Figure 4.4. The trend of the $h^+(s^+)$ correlation.

4.1.3. MINIMUM FILM THICKNESS

Under a bearing load the shaft is no longer concentric with the bearing shell. This eccentric position enables a lubricant wedge to be developed which is able to support the load. The minimum film thickness (h^+) can be difficult to measure and to define because there is no unique way of defining and measuring surface roughness. The diametrical clearance is also subject to manufacturing tolerances and wear and therefore changes in time. The minimum film thickness is therefore an approximate notion, though in the literature it is normally presented as a well defined quantity. The general trend of $h^+(s^+)$ is illustrated in Figure 4.4. The upper limit tends to $h^+ = 1$ for large values of s^+.

4.1.4. LUBRICANT FLOW RATE

In a journal bearing, lubricant is expressed from the interfacial film through the sides of the bearing and is recycled via a filter, a fluid cooler and a pump. Under steady operating conditions the replenishment cycle rate and the cooler perform-ance should be adequate to limit the lubricant temperature rise, inhibiting lubricant degradation. For a positive displacement pump driven circuit the volumetric flow rate of fluid is only slightly dependent upon the back pressure at the pump outlet and therefore the pump delivery rate is mainly dependent upon the pump speed. But

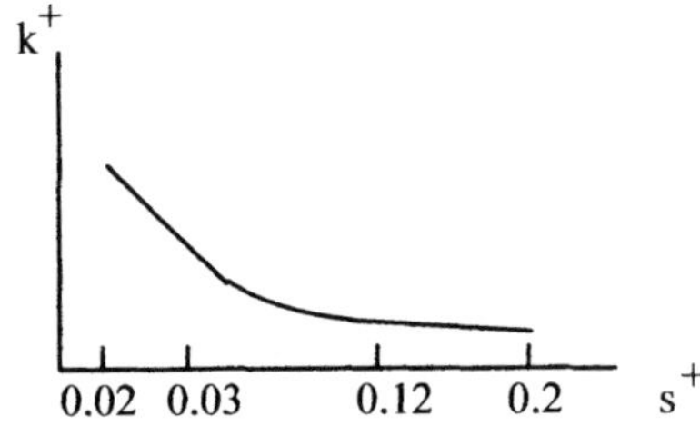

Figure 4.5. The general form of the $k^+(s^+)$ correlation.

if fluid is shared between two or more non-identical units, it would be necessary to regulate the relative distribution of lubricant between the units.

The volumetric flow rate (q) of lubricant through a journal bearing is dependent upon its delivery pressure to the bearing (p), its viscosity (η), the geometric features of the bearing (c, l, d), the specific load (w/dl) and the shaft rotational speed (n),

Thus,

$$q = q(\eta, c, d, l, p, w/dl). \tag{4.4}$$

By dimensional analysis,

$$q^+ = f((d/c)^2 \eta n/(w/ld), l/d \tag{4.5}$$

or

$$q^+ = f(s^+)(l/d)^m. \tag{4.6}$$

The value of m depends upon the fluid supply groove geometry, but it may be deduced that for an axial groove distribution, m is about 0.75. Experimental data also indicates that $f(s^+)$ is approximately

$$f(s^+) = 0.1309 + 0.1963k^{+2}, \tag{4.7}$$

where $k^+ = k^+(s^+)$.

$$\text{For } 0.02 \leq s^+ \leq 0.03 \quad k^+ = 1.036 - 18s^+,$$

$$\text{and } 0.12 \leq s^+ \leq 0.2 \quad k^+ = 0.2 - 0.6s^+.$$

The general trend of the $k^+(s^+)$ relationship is shown in Figure 4.5.

If the data for journal bearings, which is summarised in Figures 4.2 to 4.5, provided information about accuracy and scatter of the experimental data, it would be possible to formulate a corresponding fuzzified representation of the information. Alternatively, viewing the information as sufficiently precise, empirical deterministic formulae may be found for the nearly linear parts of the graphs, linked with fuzzy based transition curves to provide extensive formulae covering the

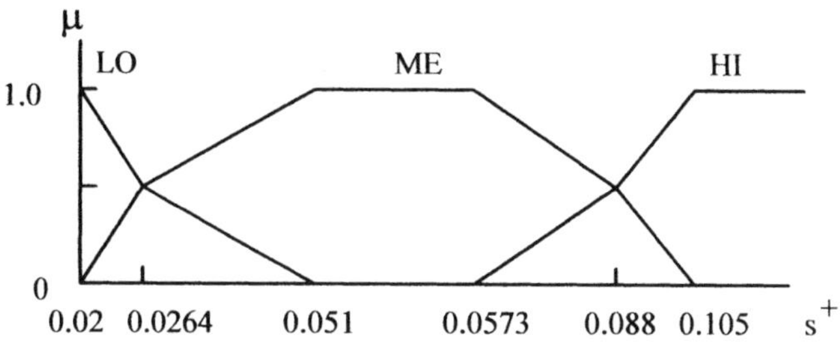

Figure 4.6. Partitioning of the s^+ UD.

whole range of the data in each chart. It is observed that this may be overstating the certainty of the relationships.

4.2. Empirical Deterministic Formulae

4.2.1. VIRTUAL FRICTION COEFFICIENT

The shape of the $\phi^+(s^+)$ curve illustrated in Figure 4.2 may be closely represented by three linear segments joined by two transition curves. The linear segments are expressed as follows:

$$\phi_{\mathrm{ME}}^+ = 13.265 + 0.55 \quad (0.051 \le s^+ \le 0.057), \tag{4.8}$$

$$\phi_{\mathrm{HI}}^+ = 6.214 s^+ + 1.17 \quad (0.105 \le s^+ \le 0.14), \tag{4.9}$$

$$\phi_{\mathrm{LO}}^+ = 0.9 \text{ constant} \quad (\text{at } s^+ = 0.02). \tag{4.10}$$

Now define ϕ by

$$\phi^+ = \mu_{\mathrm{LO}}^n \phi_{\mathrm{LO}}^+ + \mu_{\mathrm{ME}}^n \phi_{\mathrm{ME}}^+. \tag{4.11}$$

The s^+ space is partitioned as shown in Figure 4.6.

The membership functions corresponding with Figure 4.6 are listed below.

	$0.02 \le s^+ \le 0.0264$	$0.0264 \le s^+ \le 0.051$	$0.051 \le s^+ \le 0.0573$	$0.0573 \le s^+ \le 0.088$	$0.088 \le s^+ \le 0.105$
μ_{LO}	$2.563 - 78.13 S^+$	$1.0366 - 20.33 S^+$	0	0	0
μ_{ME}	$78.13 S^+ - 1.563$	$20.33 S^+ - 0.0366$	1	$1.9332 - 16.29 S^+$	$29.41 S^+ - 2.088$
μ_{HI}	0	0	0	$16.29 S^+ - 0.9332$	$3.088 - 29.41 S^+$

	$0.105 \le s^+$
μ_{ME}	0
μ_{HI}	1

Table 4.1. Sample values of l^+.

s^+	l^+
0.02	141.4
0.0264	118.0
0.035	93.17
0.51	75.54
0.57	71.97
0.070	61.35
0.088	53.73
0.105	54.51
0.120	50.16

Referring to Figure 4.2, if on $s^+ = 0.0264$, $\phi^+ = 1.0$ and $\mu = 0.5$, also $\phi_{LO}^+ = \phi_{ME}^+ = 0.9$. Inserting these values in Equation (4.11) yields $n = 0.848$. Hence,

$$\phi^+ = \mu_{LO}^{0.848}\phi_{LO}^+ + \mu_{ME}^{0.848}\phi_{ME}^+. \tag{4.12}$$

The same method may be applied to the transition curve between the two linear curves, ME and HI in Figure 4.2. Assuming that the transition curve passes through the point (0.105, 1.822), the following expression is obtained:

$$\phi^+ = \phi_{ME}^+\mu_{ME}^{1.190} + \phi_{HI}^+\mu_{HI}^{1.190}. \tag{4.13}$$

Finally, by combining Equations (4.12) and (4.13) an extensive deterministic equation is obtained:

$$\phi^+ = \mu_{LO}^{0.848}\phi_{LO}^+ + (\mu_{ME}^{0.848} + \mu_{ME}^{1.190})\phi_{ME}^+ + \mu_{HI}^{1.190}\phi_{HI}^+. \tag{4.14}$$

This equation may be used with either precise or fuzzy input data (through the extension principle).

4.2.2. FRICTIONAL POWER LOSS

The frictional power loss (l^+) is given by Equation (4.3), where ϕ^+ is obtained from Equation (4.14) above. Sample values are given in Table 4.1.

4.2.3. MINIMUM FILM THICKNESS

The minimum film thickness (h^+) is difficult to evaluate in practice even under the special conditions of the research laboratory. Surface asperities, manufacturing tolerances and load deflections may lead to significant errors in its evaluation. Confidence in a deterministic relationship is therefore lower than with other quantities.

Table 4.2. A heuristic relationship array for h^+.

L/D	LO	LM	HM	HI
DZ		HZ	HL	HH
DL		HZ	HM	HH
DM		HL	HM	HV
DH		HM	HH	
DV	HZ	HM	HV	

(Column group header over LO, LM, HM, HI is **log s^+**.)

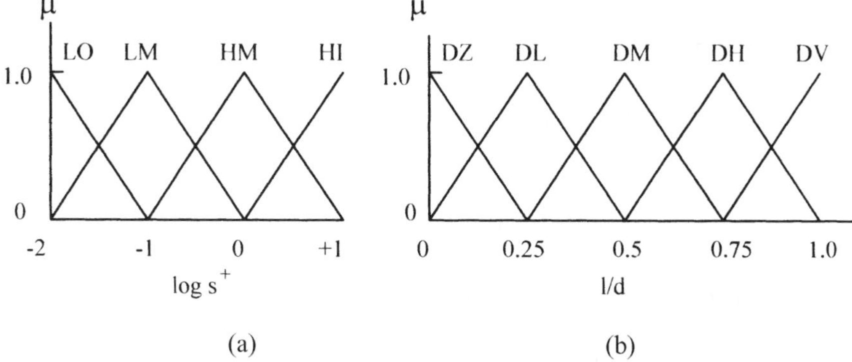

Figure 4.7. Partitioning of the journal bearing parameter UDs. (a) $\log s^+$. (b) l/d.

A heuristic relationship which does not require mathematical formulation is more appropriate. Published data indicates that $\log s^+$, rather than s^+ is more appropriate as an independent variable because of the range of s^+ compared with h^+.

For a given value of s^+, h^+ is also dependent upon the length to diameter ratio of the bearing (l/d). From the available published data a heuristic relationship such as that given in Table 4.2 may be proposed.

The partitioning of the $\log s^+$, l/d and h^+ UDs is shown in Figure 4.7. A relatively coarse partitioning is chosen reflecting the level of knowledge about the detail.

Relative to h^+, the values of l/d and $\log s^+$ would be precise and could be represented by simple numbers.

An example of finding the minimum film thickness using the above method is given below.

EXAMPLE 4.1. Under service conditions the Sommerfeld number for an oil lubricated journal bearing is estimated to be 0.471 and its design length/diameter ratio

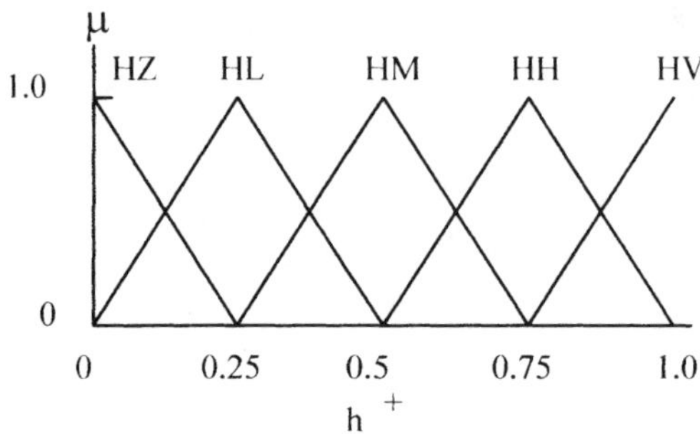

Figure 4.8. Partitioning of the h^+ UD.

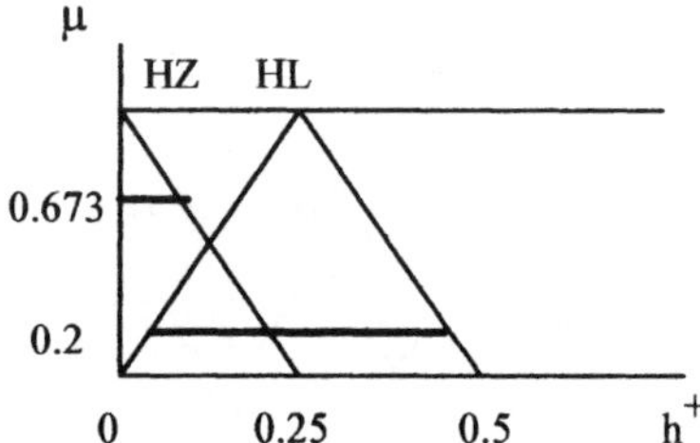

Figure Ex 4.1. Journal bearing fuzzy conclusion.

is 0.3. Evaluate the nominal minimum oil film thickness.

Solution. If $s^+ = 0.0471$, then $\log s^+ = -1.327$.

Assuming that Figures 4.7(a), 4.7(b) and 4.8 define the partitioning, the following membership values may be found:

Figure 4.7(a) $\mu_{\text{LO}} = -(-1.327 + 1) = 0.327$ and $\mu_{\text{LM}} = 1 - 0.327 = 0.673$,

Figure 4.7(b) $\mu_{\text{DM}} = (0.3 - 0.25) = 0.2$ and $\mu_{\text{DL}} = 1 - 0.2 = 0.8$.

The FL proposition is

IF LOG S$^+$	AND L/D	THEN H$^+$	MIN	CONSEQUENCE
LO	DM	–		
LO	DL	–		
LM	DM	HL	0.673, 0.2	0.2HL
LM	DL	HZ	0.673, 0.8	0.673HZ

The conclusion is $h^+ = 0.673\text{HZ} \cup 0.2\text{HL}$. This conclusion is illustrated in Figure Ex 4.1.

It may easily be shown that for membership values of 0.2 and 0.673, the values of $H+$ are 0.05 and 0.0818 respectively. Thus from Equations (1.9) and (1.10) the EFN may be found.

$$w = (z - x)/(2 - m - n) \tag{1.9}$$

and

$$c = x + w(1 - n). \tag{1.10}$$

The values to be used in these equations are: $m = 0.673$, $n = 0.2$, $x = 0.05$ and $z = 0.0818$. Hence, $w = 0.0724$ and $c = 0.108$. The value of c may used as a characteristic number.

It may be noted from Figure Ex 4.1 that the dominant part of the solution lies in the lowest fuzzy set (HZ), which ranges from $h^+ = 0$ to $h^+ = 0.25$. In practice it may be desirable to use a higher viscosity oil or increase the oil cooling rate to reduce its temperature and therefore its viscosity (or both). This would introduce a stronger influence of the HL fuzzy set on the conclusion. At the design stage there is the option of increasing the l/d ratio. Such a condition as that found above would suggest that in a real plant it should be closely monitored during the plant commissioning stage and on a regular basis thereafter with a view to adjusting the operating conditions. The oil cooler capacity should have sufficient capacity and controllability to enable the oil temperature to be adjusted independently of its flow rate to the bearing.

If the nominal bearing diametrical clearance was 200 microns, then the nominal minimum oil film thickness, based upon the principal value of the EFN (0.108) would be 10.8 microns.

4.2.4. LUBRICANT FLOW RATE

The basic form of the lubricant flow rate relation has been discussed in Section 4.1.4. It comprised two near linear portions as reflected in Figure 4.5. A membership function approach may now be applied to find a transition expression between these two branches.

Let the membership functions have the form

$$\mu_1 = a \exp -bs^+ \quad \text{and} \quad \mu_2 = 1 - \mu_1$$

with boundary conditions $\mu_1 = 0.01$ on $s^+ = 0.12$ and $\mu_1 = 1.0$ on $s^+ = 0.02$.

Applying the boundary conditions to the membership function yields

$$\mu_1 = 2.512 \exp -46.05s^+.$$

Now,

$$f(s^+) = 0.1309 + 0.1963k^{+2}, \tag{4.7}$$

where k^+ is given by $k^+ = \mu_1 k_1^+ + \mu_2 k_2^+$.

$$\text{For } 0.02 \le s^+ \le 0.03, \quad k_1^+ = 1.036 - 18s^+,$$

$$\text{and } 0.12 \le s^+ \le 0.2, \quad k_2^+ = 0.2 - 0.6s^+.$$

Table 4.3. Values of the dimensionless lubricant volumetric flow rate.

s^+	q^+	s^+	q^+
0.03	0.179	0.08	0.136
0.04	0.141	0.09	0.136
0.05	0.136	0.10	0.135
0.06	0.135		
0.07	0.136		

Tabulated values of the values of the lubricant flow rate (q^+) for an l/d ratio of unity are given in Table 4.3.

The lubricant flow rate falls more rapidly at first, then levels to a more-or-less constant value as the Sommerfeld number increases and the eccentricity becomes small.

Under steady state conditions, and neglecting radiant and convective cooling, the lubricant temperature rise of the lubricant passing through the bearing is due to the friction heating. A heat balance gives

$$l = \rho \alpha q \Delta t, \tag{4.15}$$

where ρ is the lubricant density, α is its specific heat and Δt is its temperature rise.

4.3. Fuzzy Input Data

The usual case in the published literature is that of precise input data applied to deterministic relations, which of course gives precise solutions (conclusions). In all the relationships in this chapter, whether they are deterministic or fuzzy, either precise or fuzzy input data can be applied. In both these cases the conclusion is fuzzy.

Consider the case where the relationships for a system are derived from precise research data. Service data is obtained which lacks the precision of the research data and is available in the form of fuzzy numbers. The following example illustrates the result of uncertainty in the service viscosity and the bearing diametrical clearance.

EXAMPLE 4.2. An oil lubricated journal bearing has a shaft diameter of 0.25 m and its diametrical clearance in the bearing at the half-service life stage is approximately 200 microns. The steady shaft load is 38 kN and its running speed is 400 rev/min. The bearing length is 0.075 m. The bearing operates at a temperature of about 50°, at which temperature the oil viscosity would be 0.037 Pa s. The error in the diametrical clearance is estimated to be ±10%, whilst the uncertainty in the interfacial oil viscosity and the variability due to ambient conditions contribute to an overall uncertainty of $\pm 20\%$.

Find the corresponding virtual coefficient of friction.

Solution. In the following solution the piecewise continuous fuzzy values will be represented as discrete fuzzy factors of the nominal values for ease of manipulation. Thus, if η_0 and c_0 are the nominal viscosity and diametrical clearance values, the fuzzy data values are

$$\eta = (0//0.8 + 1//0.9 + 1//1.1 + 0//1.2)\eta_0.$$

The fractional numbers in this expression are all multipliers for μ_0. Also,

$$c = (0//0.9 + 1//1 + 0//1.1)c_0.$$

Hence,

$$d/c = (0//1.11 + 1//1 + 0//0.909)(d/c_0).$$

Taking the Cartesian product of the membership values and the algebraic products of d/c yields

$$(d/c)^2 = (0//1.23 + 1//1 + 0//0.827)(d/c_0)^2.$$

Now forming the Cartesian product array with the viscosity,

	0.8	0.9	1.1	1.2
1.23	0	**0**	0	0
1.0	**0**	1	1	0
0.827	0	0	0	0

$$\eta(d/c)^2 =$$ (above) $$//$$

0.984	**1.107**	1.353	1.476
0.800	**0.900**	**1.100**	1.200
0.662	0.744	0.910	0.992

$$\eta_0(d/c_0)^2.$$

The principal values in the above factor arrays are shown in bold type. These values may be used to factor the nominal Sommerfeld number to find its spread. The distribution provided by these values is illustrated in piecewise continuous form in Figure Ex 4.2.

The Sommerfeld number for a bearing with $l/d = 1$ may be calculated from the given data as $s_1^+ = 0.1903$. The present bearing has an $l/d = 0.3$ and the nominal Sommerfeld number (s_0^+) for value is

$$s_0^+ = s_1^+(l/d)^{1.16} = 0.1903 * 0.3^{1.16} = 0.0471.$$

The $\mu(d/c)^2$ arithmetic factors in the above array may be applied to the nominal Sommerfeld number (s_0^+). Only the principal values are required to portray the resulting distribution. The resulting distribution of the Sommerfeld number (s^+) is

–	0.0521	–
0.0377	0.0424	0.0518
–	–	–

$$s^+ =$$ (above)

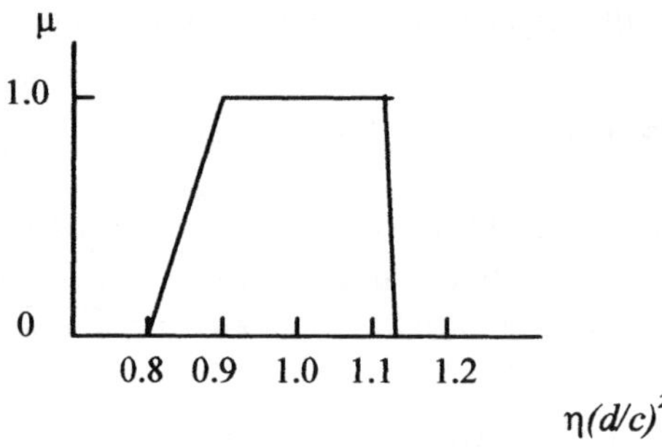

Figure Ex 4.2. The fuzzy factor distribution.

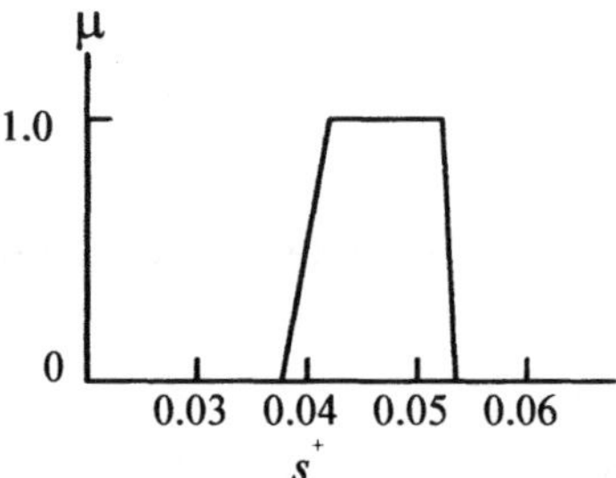

Figure Ex 4.3. The fuzzy Sommerfeld distribution.

This array is illustrated as a piecewise continuous function in Figure Ex 4.3.

Considering each value of the fuzzy Sommerfeld number in turn and substituting as appropriate into the equations given in Section 4.2.1 provides the results in Table Ex 4.1.

Table Ex 4.1. Values of the fuzzy virtual coefficient of friction factors.

s^+	ϕ^+
0.0377	1.121
0.0424	1.149
0.0518	1.237
0.0521	1.243

The values given above are, of course, subject to the membership values given in the array.

The conclusion is therefore:

$$\phi^+ = [0//1.121 + 1//1.149 + 1//1.237 + 0//1.243].$$

4.4. Remarks

The published hydrodynamic lubrication correlations are the results of theoretical and experimental research and are semi-theoretical in nature. The effects of scatter and errors in the experimental results are generally not reported. The outcome of the work is reported as deterministic correlations between various parameters such as; friction coefficient, power loss, minimum film thickness and lubricant flow rate.

In this work empirical representations of the parameters have been patched together to form extensive equations using the concepts of membership functions. The exception is the minimum film thickness which has been treated by a fuzzy relational array. This parameter can be of the order of a few tens of microns and is difficult to measure. It is therefore more prone to experimental error. Manufacturing geometric and surface tolerances may have a noticeable effect on its value, hence a deterministic relationship is not so plausible.

Service conditions are invariably poorly defined compared with those of the research laboratory and the uncertainty in their values should find more realistic expression in fuzzy numbers, rather than in precise values, even if these are qualified by the term "approximately". The deterministic relations may still be used (if they are plausible) in conjunction with the extension principle. The conclusions then convey a clearer idea of the possibilities.

No explicit recognition of cavitation is given in the data used in this work. It is a significant phenomenon in practice and may be associated with foaming of the lubricant. Within the interface of the bearing, down-stream of the minimum film thickness the film pressure may fall below the ambient value and this introduces air entrainment into the lubricant film. This can be associated with lubricant starvation and bearing overheating. Obviously, the lubricant supply point should be in the low pressure region of the bearing. At the design stage, cavitation conditions can be deterred by increasing the bearing length, or by decreasing the load. In service, increasing the lubricant flow rate, or changing its grade to increase its viscosity would normally help.

Chapter 5
Elastohydrodynamic Lubrication

In lubrication practice it has been found that some non-conformal rubbing metal surfaces are able to perform satisfactorily under service conditions that apparently precluded film lubrication. Important examples being found in gear trains and roller bearings. The phenomenon remained unexplained until the combined effects of lubricant compressibility and elastic deformation of the bearing surface were taken into consideration. The former effect had infact been suggested in various earlier studies as an explanation of the "oiliness" of lubricants. The combined effects have explained the fundamental reasons for the satisfactory performance of gears and also ball and roller bearings under load. It may be mentioned that natural (animal) joints, where lubricant compressibility is not a factor, benefit from compliant bearing material (cartilage) and perform remarkably well in the healthy state. The general term for the phenomenon in engineering is Elastohydrodynamic Lubrication (EHL).

The EHL treatment applies in practice where there is a low degree of conformity between the rubbing surfaces accompanied by high contact pressures. In extreme circumstances, where the surfaces are separated only by adsorbed molecular surface films, or are in incipient metal-to-metal contact, such as occurs during machinery start-up, the problem lies in the field of surface chemistry and physics rather than fluid dynamics. A major concern in EHL studies is that of the lubricant film thickness at the contact interface.

5.1. Deterministic Analysis

5.1.1. FILM THICKNESS

Both the theoretical and experimental EHL research conditions are usually rather different to those prevailing in practice. The former are based upon steady state conditions, whereas the latter are normally highly transient in nature. For example, laboratory two and four-ball machines are run at a steady speed and load, whereas in a spur gear the tooth contact load is not steady and neither is the rubbing/sliding speed combination steady. Also in the theory the isothermal rather than the adiabatic lubricant compressibility is used. In service conditions, vibrations, malalignment, load deflections, manufacturing tolerances, surface texture and wear are also effects that distinguish the in-service conditions from those of research work, where the objective is to make the experimental conditions approach those of the theoretical model as closely as possible. Errors of observation and the repeatability of experimental results are frequently not disclosed in published work and the precision and robustness of the resulting physical relationships cannot be asserted in this case.

Satisfactory lubrication of gears and ball or roller bearings is contingent upon the existence and maintenance of a sufficient oil film in the contact zone interface under service conditions. Dowson and Higginson (see Further Reading) have reported extensive theoretical and experimental work aimed at providing semi-theoretical deterministic steady state equations for the hydrodynamic and the EHL contact zones. The results are presented as relationships between the dimensionless sliding speed (u^+), the load (w^+) and the film thickness (h^+). The HL and EHL zones are bridged by an unspecified transition zone.

In the HL zone the relationship is expressed as

$$h^+ = 4.9u^+/w^+, \tag{5.1}$$

where $h^+ = $ (film thickness)$/r$ and r is an equivalent contact radius, $u^+ = $ (lub. visc)$*u/E'r$ and $w^+ = w/E'r$. u is the mean speed and w is the actual surface load. E' is the effective Young's modulus of the metal surfaces.

For the EHL zone, the relationship is

$$h^+ = 1.6G^{0.6}u^{+0.7}/w^{+0.13}, \tag{5.2}$$

where $G = \alpha E'$ and α is the (isothermal) pressure exponent of viscosity. For a steel/mineral oil combination G is approximately 5000. Thus, for this case,

$$h^+ = 256.2u^{+0.7}/w^{+0.13}. \tag{5.3}$$

Hence, in the HL zone h^+ is inversely proportional to w^+, but in the EHL zone it is only a weak function of w^+ according to these models. These relationships are presented in the form of a chart of $\log u^+$ as a function of $\log w^+$ with h^+ as a parameter. This is illustrated in Figure 5.1.

5.1.2. THE TRANSITION ZONE

The transition zone between the HL and the EHL zones is a mixture of both types of lubrication in varying degrees. This may be treated on a fuzzy membership basis. Let Equation (5.1) be rearranged as

$$u_1^+ = 0.2041h^+w^+ \tag{5.4}$$

and Equation (5.3) as

$$u_2^+ = 2899h^{+1.429}w^{+0.1857}. \tag{5.5}$$

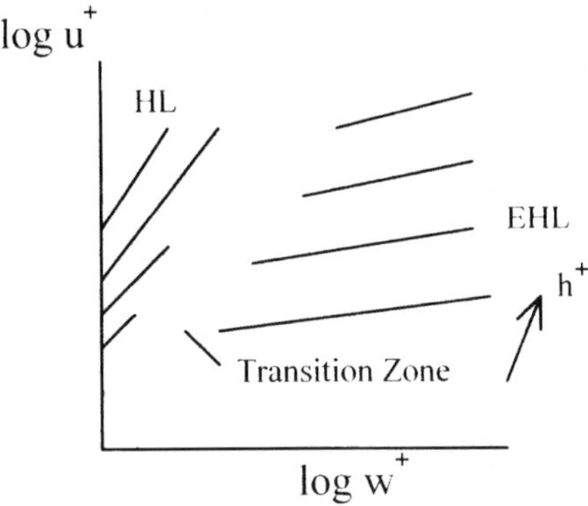

Figure 5.1. An illustration of the lubrication zones.

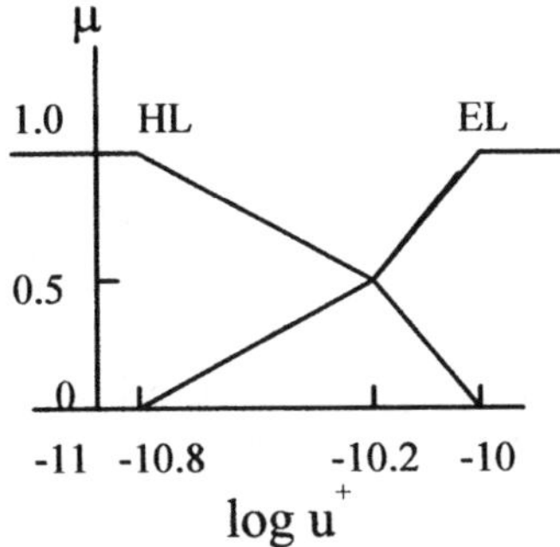

Figure 5.2. Partitioning of the $\log u^+$ UD.

Now finding the intersection of u_1^+ and u_2^+ with a mid-range trial value of $h^+ = 10^{-4}$,

$$2.041 * 10^{-5} w^+ = 6.962 * 10^{-10} w^{+0.1857}.$$

Hence,

$$w^+ = 3.67 * 10^{-6}. \tag{5.6}$$

Then at the intersection,

$$u_1^+ = u_2^+ = 6.7 * 10^{-11}. \tag{5.7}$$

The chart value is in the region of $8 * 10^{-11}$ at the intersection point. The limits of the transition zone for $h^+ = 10^{-4}$ are $\log u^+ = -10.8$ and -10. The $\log u^+$ UD is judged to be conveniently partitioned as shown in Figure 5.2.

The membership functions associated with the partitioning in Figure 5.2 are given below.

	$\log u^+ \le 10.8$	$10.8 \le \log u^+ \le -10.2$	$-10.2 \le \log u^+ \le -10$	$-10 \le \log u^+$
μ_{HL}	1	$-8 - 0.8333 \log u^+$	$-25 - 2.5 \log u^+$	0
μ_{EL}	0	$0.8333 \log u^+ + 9$	$-2.5 \log u^+ + 26$	1

$$u^+ = \mu_1^n u_1^+ + \mu_2^n u_2^+. \tag{5.8}$$

The curve for $h^+ = 10^{-4}$ passes through $u^+ = 6 * 10^{-11}$ on $w^+ = 3.282 * 10^{-6}$. Hence from Equation (5.8),

$$6 * 10^{-11} = 2 * 0.5^n * 6.7 * 10^{-11},$$

which gives

$$n = 1.16. \tag{5.9}$$

Therefore, Equation (5.8) becomes

$$u^+ = \mu_1^{1.16} u_1^+ + \mu_2^{1.16} u_2^+. \tag{5.10}$$

Equation (5.10) together with the above membership functions span the whole range of physical conditions; HL, transition and EHL alike. Within the accuracy of the basic data. But there is no information about the experimental errors; therefore the equations may need to be interpreted as deterministic, but with fuzzy coefficients and indices. In fact, there are some discrepancies in the experimental data indicating that the index on h^+ in Equation (5.5) may be nearer to 2.

5.2. Application to Service Conditions

The form of Equations (5.4), (5.5) and (5.10) may be used as a guide to the formulation of fuzzy relations and suggests that appropriate UDs are $\log w^+$, $\log u^+$ and $\log h^+$. Partitioning of these UDs based upon a judgement of the precision of research findings, the similarity between the laboratory and real service conditions (spur gears) and the accuracy of service data, is shown in Figures 5.3, 5.4 and 5.5.

The relationship array corresponding with Figures 5.3, 5.4 and 5.5 is shown in Table 5.1.

The above partitioning and relationship array describes the lubrication of nonconforming surfaces in general terms. The description has been guided by the semi-theoretical results of research studies on idealised conditions intended to be homologous to actual service conditions. The nearest practical cases are those of the roller bearing and straight spur gear.

In some cases the available design or in-service data is cast in terms of precise quantities which need to be translated into terms compatible with the corresponding fuzzy relationship array. The fuzzification process follows the usual pattern.

EXAMPLE 5.1. In a preliminary straight spur gear design study, it is estimated that the nominal tooth load is likely to be about $w^+ = 5.4 * 10^{-4}$ and the nominal speed

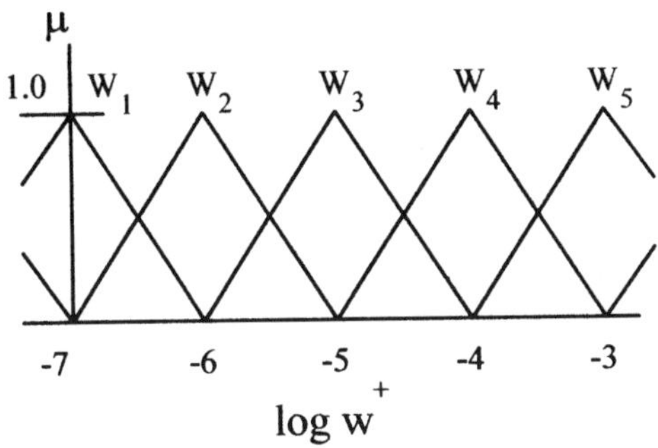

Figure 5.3. Partitioning of the $\log w^+$ UD.

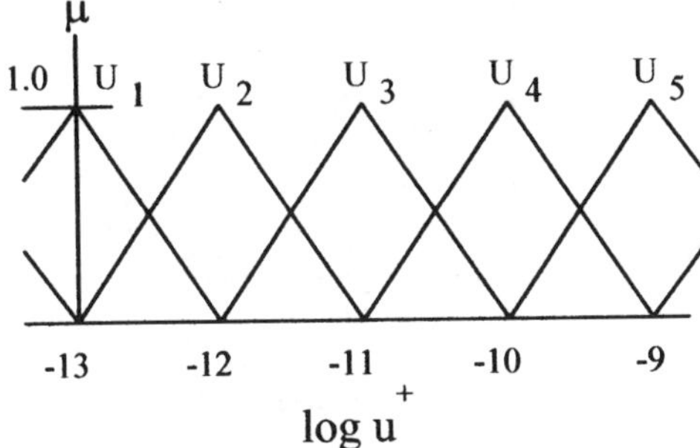

Figure 5.4. Partitioning of the $\log u^+$ UD.

$u^+ = 8.15 * 10^{-11}$. Also assume that $G = 5000$. Find the dimensionless nominal film thickness between the gear teeth.

Solution. Assuming the partitioning of the parameters as shown in Figures 5.3, 5.4 and 5.5 and also the relationship array in Table 5.1.

From the given data: $\log w^+ = -3.267$ and $\log u^+ = -10.09$.

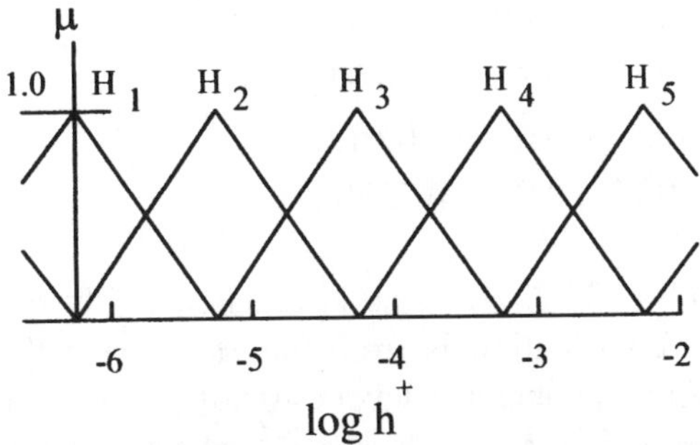

Figure 5.5. Partitioning of the $\log h^+$ UD.

Table 5.1. Relationship array for h^+.

	W				
U	W_1	W_2	W_3	W_4	W_5
U_1		H_1	H_1	H_1	
U_2	H_3	H_2	H_2	H_2	
U_3	H_4	H_3	H_3	H_2	H_2
U_4	H_5	H_4	H_3	H_3	H_2
U_5		H_5	H_4		

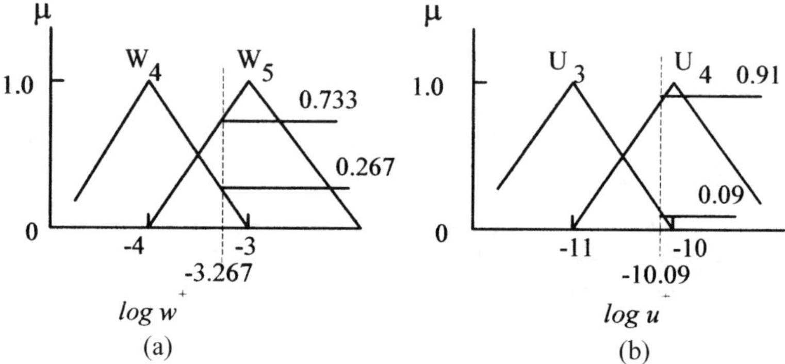

Figure Ex 5.1. Fuzzification of the gear data.

From Figure Ex 5.1, membership values of the gear data are

$\log w^+$: $\mu_{W4} = 0.267$ and $\mu_{W5} = 0.733$,

$\log u^+$: $\mu_{U3} = 0.09$ and $\mu_{U4} = 0.91$.

The fuzzy proposition is:

IF W	AND U	THEN H	MIN	CONSEQUENCE
W_4	U_3	H_2	0.267, 0.09	$0.09H_2$
W_4	U_4	H_3	0.267, 0.91	$0.267H_3$
W_5	U_3	H_2	0.733, 0.09	$0.09H_2$
W_5	U_4	H_3	0.733, 0.91	$0.733H_3$

The conclusion is $H = 0.09H_2 \cup 0.733H_3$. This conclusion is illustrated in Figure Ex 5.2. Clearly, the major contribution is from the H_3 set, which has a principal value ($\mu = 1$) of -4.1. At a membership level of 0.5 the range of $\log h^+$ is $-4.6 \le \log h^+ \le -3.6$, or $2.51*10^{-5} \le h^+ \le 25.1*10^{-5}$. The designer should recognise this as a possible range of values.

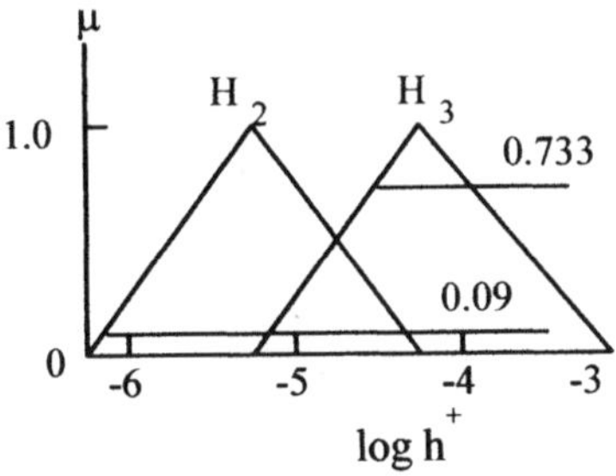

Figure Ex 5.2. The fuzzy conclusion for h.

The equivalent fuzzy number to the conclusion may be found by applying Equations (1.9) and (1.10). Thus for the membership values of $m = 0.733$ and $n = 0.09$, the corresponding abscissa values are $x = -4.367$ and $z = -4.19$, respectively.

Now,

$$w = (z - x)/(2 - m - n) \qquad (1.9)$$

and

$$c = x + w(1 - n). \qquad (1.10)$$

Hence, $a = -4.381$, $b = -4.081$ and $c = -4.231$. The equivalent fuzzy number as a discrete set is

$$\text{EFN}(H) = [0// - 4.381 + 0.5// - 4.306 + 1// - 4.231 +$$
$$+ 0.5// - 4.156 + 0// - 4.081].$$

The principal value of this fuzzy set is -4.321.

For comparison the fuzzy conclusion may be defuzzified by the weighted abscissa method (see Equation (1.5)). This gives $\text{Def} \log h^+ = -4.204$.

5.3. Fuzzy Inputs

It is unlikely that precise values of elements of W^+ and U^+ would be known under service conditions. The elements of U^+ depend upon the lubricant viscosity, which in turn depends upon its temperature. The values of W^+ are not constant but fluctuate throughout the tooth contact cycle. There are also imperfections mentioned earlier some of which are difficult to quantify, but which add to the uncertainty. This means that W^+ and U^+ values will generally both need to be expressed as fuzzy numbers. To see the effect of this in more detail, consider again the above example, but now with fuzzy inputs.

Suppose that the range of uncertainty in w^+ and u^+ is each equal to $\pm 15\%$ of the nominal value. The fuzzy input sets in the above example then are expressed as discrete fuzzy sets

$$W^+ = 0// - 3.338 + 1// - 3.268 + 0// - 3.207 \qquad (5.12)$$

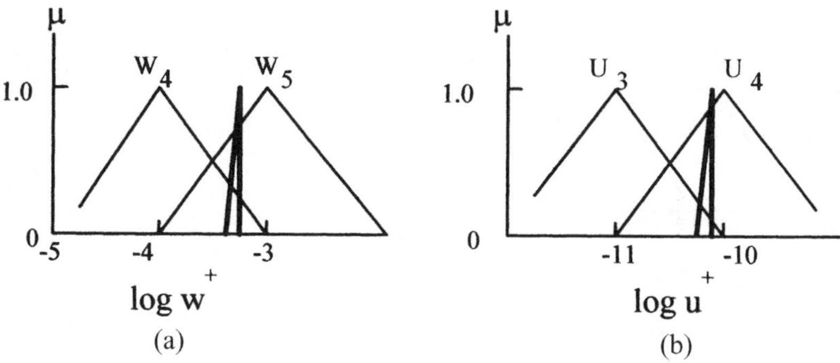

Figure 5.6. Fuzzy input sets on the $\log w^+$ and $\log u^+$ UDs. (a) Set w^+. (b) Set u^+.

and

$$U^+ = 0// - 10.16 + 1// - 10.09 + 0// - 10.03. \tag{5.13}$$

These inputs are shown as piecewise continuous sets on the $\log w^+$ and $\log u + $UDs in Figures 5.6(a) and (b) respectively.

It will be noted that there are pairs of intercepts of the inputs on both the $\log w^+$ and $\log u^+$ UDs. This means that there are four possible conclusions of $\log h^+$. The conclusions may be found by the same process as before. In terms of membership values, the intercepts are

$$\log W^+: \quad \mu_{MW} = 0.733, \quad \mu_{PW} = 0.769,$$

$$\mu_{NW} = 0.231, \quad \mu_{QW} = 0.308,$$

$$\log U^+: \quad \mu_{MU} = 0.885, \quad \mu_{PU} = 0.923,$$

$$\mu_{NU} = 0.038, \quad \mu_{QU} = 0.192.$$

Applying the same methods as in the above example and considering the membership value pairs (μ_{MW}, μ_{NW}) and (μ_{MU}, μ_{NU}). The two conclusions are

$$H_a = 0.09 H_2 \cup 0.733 H_3 \tag{5.14}$$

and

$$H_b = 0.192 H_2 \cup 0.763 H_3. \tag{5.15}$$

Also, for the membership value pairs (μ_{PW}, μ_{QW}) and (μ_{PU}, μ_{QU}) the conclusions are

$$H_c = 0.038 H_2 \cup 0.769 H_3, \tag{5.16}$$

$$H_d = 0.192 H_2 \cup 0.885 H_3. \tag{5.17}$$

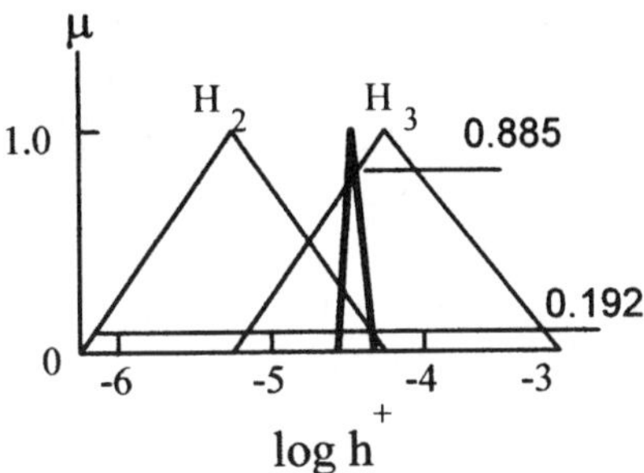

Figure 5.7. Fuzzy conclusion and the EFN.

The resultant conclusion is the union of the above conclusions in Equations (5.14) to (5.17),

$$H = H_a \cup H_b \cup H_c \cup H_d \tag{5.18}$$

$$= 0.192 H_b \cup 0.885 H_c. \tag{5.19}$$

The EFN is given by $a = -4.312$, $b = -4.225$, $c = -4.038$.

The resultant fuzzy conclusion, Equation (5.18) and the EFN are illustrated in Figure 5.7.

5.4. Remarks

Even if the published relationships of EHL were precise and the experimental data from two and four ball laboratory tests agreed with negligible error, there would still be a need to translate the relationships into FL form for practical application. The reason for this is the difference between the research laboratory conditions and those found in real applications. (This is usually greater in the case of EHL than HL.) There is also the usual lack of precision of knowledge of the in-service data. The FL treatment is particularly appropriate in such circumstances.

The partitioning of the UDs reflects the confidence in the accuracy and reproducibility of the research data. The relationship array is the key to bridging the difference between the research results and reality. It fuses the results with experience and judgement that is otherwise difficult or impossible to achieve with deterministic formulae. Moreover, the array can be readily modified in the light of feedback of further service experience. It may also be modified to extend the application to other types of geometry. For example, a straight spur gear relationship array may be modified for application to helical gearing on the basis of practical observation and experience. The convenient merging of research results and in-service experience is unmatched by any other type of treatment.

In the conventional design practice of applying deterministic formulae (which requires precise input data), the solutions to problems are given in terms of precise numbers, sometimes with the caveat, "approximately". In contrast, the FL

                    Fuzzy Logic Applications in Engineering Science

treatment provides a fuzzy (usually sub-normal) set for the conclusion. This gives a continuous range of values (within limits) with varying degrees for weighting. The designer must interpret this by deciding on the cut-off level of weighting that satisfies the acceptable risk level of reliability. There is a tendency for the result to cluster around a value which may be determined by defuzzifying the conclusion, but this cannot be taken as the sole and final solution to a problem because there is a loss of information and there is risk in the full conclusion.

Chapter 6
Fatigue and Creep

Fatigue and creep are two frequently experienced causes of failure in the operation of systems such as industrial plant, transport vehicles, railways and aircraft. The phenomena imply accumulated material damage, often as a result of service loads and/or elevated temperatures. Common engineering solids, including metals and plastics, are susceptible to these forms of damage with a concomitant reduction in remnant life span at any point in the service life of the system. The possibility of such damage requires in-service inspection and management of maintenance action to arrest deterioration.

At the design stage, the available information relating to fatigue and creep may comprise some or all of the following:

(a) forecast load patterns, residual stresses, thermal and other environmental information relating to in-service conditions; also the component microstructure;

(b) laboratory model and sample material properties diagnostic test data with the same microstructure;

(c) knowledge and experience of the performance and problems of similar systems;

(d) specification of the required system performance and reliability;

(e) specification of the acceptable risk levels.

The nature of fatigue loading is very varied in practice, from the tens of millions per annum of automobile components at the high endurance end of the fatigue spectrum, where only small elastic strains are permissible, to that of large systems such as chemical plant and aircraft which experience a few to the low hundreds of cycles per annum. The low endurance end of the fatigue spectrum is different in that stressing in the plastic range of the material is permissible and the system may have a design service life of perhaps 20 years, representing an extensive reliable and economic service life.

Creep is more often associated with low-endurance fatigue in which there may be dwell periods. Turbine blades are rather different in that creep due to centrifugal effects is associated with high-frequency fluctuating bending stresses due to stator/rotor interactions.

Large structures pose a particularly difficult class of problems in that both the physical and time scales of the real cases precludes close replication in the laboratory. There is also the problem of scale effects. Prototype proving tests may not be possible, especially if they are a one-of-a-kind design. Under these circumstances recourse must be made to model and material specimen test with the knowledge of the risks involved. It is customary then to take defensive action in generous factors of safety. The application of probability theory may enable factors of safety to be reduced, but the theory depends upon the validity of the Gaussian (normal) probability distribution function, which may or may not be appropriate. Extreme events are unlikely in this theory, which requires a stationary system and environment. There is an unexpected nature to some failure events in practice that are not predicted by current probabilistic theory.

6.1. Deterministic Analysis

6.1.1. BASIC EQUATIONS

Reliability predictions are usually based upon material and system models of the deterministic type. This entails modelling with sufficient disposable constants to enable fitting of the available experimental data to be achieved. The most common correlation in fatigue studies is the well-known $s_a - n$ high-endurance data from material specimen tests. These are conventionally fitted by an empirical formula

$$s_a n^p = a, \tag{6.1}$$

where s_a is the stress amplitude and n is the number of cycles to failure, a and p are disposable constants. In some cases where the experimental data appears better fitted by a bilinear form, then two sets of constants are required for the two branches of data.

Laboratory test data for low-endurance fatigue is often reported in terms of plastic strain (ε_p) rather than stress. In this case an alternative empirical formula is applied:

$$\varepsilon_p n^q = b, \tag{6.2}$$

where b and q are other disposable constants. There is inevitably scatter in the experimental data for both low and high-endurance fatigue tests which are not reflected in the precise forms of Equations (6.1) and (6.2), and which in reality provides a significant level of uncertainty.

Detailed examination of metallic components subject to fatigue or creep, by visual, ultrasonic and other methods of examination, reveals the presence of fine cracks within the material which grow in size as the number of cycles, or time in the case of creep, increases. The cracks are initially of microscopic scale, typically

10^{-6} to 10^{-5} m (stage I). It is considered that stage II growth results from stress concentration at the crack tip; it is observed to be a relatively slow and intermittent process at this stage and requires the bulk of the required time for rupture, which occurs during the fast crack growth (stage III) period. At stage II the cracks are of the same order of magnitude as the grain size. Stage I is intracrystalline whereas stage II is intercrystalline and grain size and shape are important.

At this scale the theoretical analysis is based upon fracture mechanics concepts. The treatment assumes a homogeneous and isotropic continuum with bulk properties rather than a grainy, inhomogeneous and possibly anisotropic material with inclusions (especially around welded joints). This represents a gulf between reality and its theoretical portrayal which is difficult to bridge using conventional methods of analysis. At the microscopic level, the concept of a smoothly varying stress field is not to be interpreted literally. Again, stresses cannot be measured directly, they are inferred from measured surface strains, perhaps by interpolation or extrapolation, and assumed material properties.

6.1.2. FRACTURE MECHANICS

The fracture mechanics studies have a point of departure in the Paris–Erdogan formula which relates the crack growth per cycle to the stress intensity factor Δk

$$\mathrm{d}a/\mathrm{d}n = G(\Delta k)^n, \tag{6.3}$$

where a is the crack length.

Finite elements are also sometimes applied to determine the stress field at the crack tip based upon similar continuum assumptions. The fracture mechanics representation is not considered further in this work. Its application is not as well established in practice as the empirical deterministic method. It aims to start from a more fundamental basis and by a process of integration arrives at a formula which is compatible with the empirical deterministic formula, Equation (6.1). Considering the objective of formulating a FL approach it offers no advantages.

6.1.3. s–n INTERSECTIONS

High endurance fatigue laboratory data sometimes correlates on a semi-log or a log-log graph, but with two branches. A typical example is shown in Figure 6.1. This is another example where a transition zone may be found using membership functions, similar to the applications in Chapters 4 and 5.

Although there is scatter in the original data, the trend of the data of the aluminium alloy in Figure 6.1 is in favour of bilinear graphs on a s–$\log n$ graph, which merge in a transition zone. The two linear plots are fitted by mean lines of the following form

$$s_1 = 853.42 - 121.48 \log n \ \mathrm{MN/m^2} \tag{6.4}$$

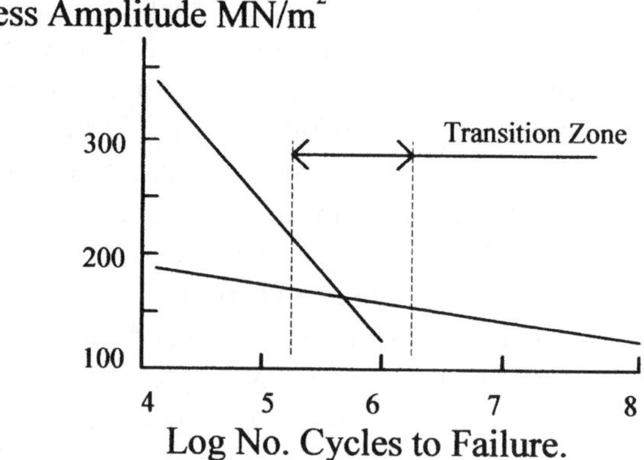

Figure 6.1. Aluminium alloy fatigue chart (after Benham and Crawford).

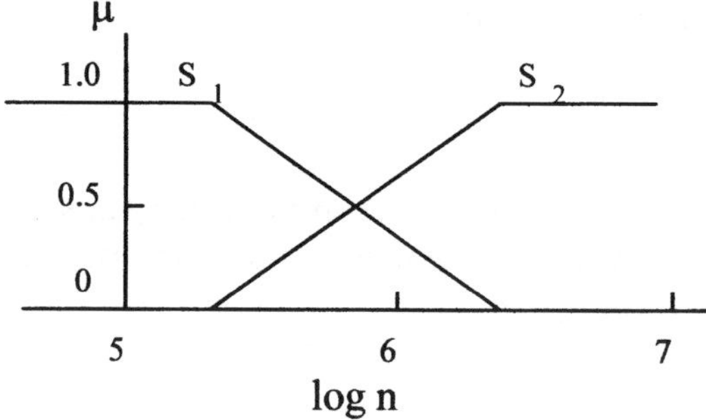

Figure 6.2. Partitioning of the $\log n$ UD for stress in the transition zone.

and

$$s_2 = 267.00 - 17.75 \log n \text{ MN/m}^2. \tag{6.5}$$

The transition zone is defined in the range; $2*10^5 \leq \log n \leq 30*10^5$ cycles. At the extremes of the range the bilinear curves are tangential to the transition curve. The two Equations (6.4) and (6.5) are considered as representing two interpenetrating fuzzy sets on the $\log n$ UD with membership functions that define the shape of the transition curve. The expression for the stress amplitude is given by

$$s = \mu_1^b s_1 + \mu_2^b s_2, \tag{6.6}$$

where the index b is a hedging index. The partitioning of the $\log n$ UD in accordance with the above conditions is illustrated in Figure 6.2.

The membership functions are

$$\mu_{s_1} = 5.508 - 0.8503 \log n \quad \text{and} \quad \mu_{s_2} = 0.8503 \log n - 4.508.$$

Table 6.1. Stress amplitude values in the transition zone.

$n * 10^{-5}$ cycles	s MN/m^2
2	209
4	183
6	170
10	155
30	152

The bilinear curves intersect at $\log n = 5.653$ and at this point a fair point on the transition curve is judged to be 179 MN/m^2. From Equation (6.6) the index b is then given as 0.9. The extensive empirical equation for stress is therefore

$$s = \mu_1^{0.9} s_1 + \mu_2^{0.9} s_2. \tag{6.7}$$

Sample values of stress amplitude in the transition zone, given by Equation (6.7) are listed in Table 6.1.

6.1.4. COMBINED STRESS ENDURANCE LIMIT

Quite often in practice a fluctuating stress is accompanied by a steady stress of the same type. This case has been studied extensively and the method of plotting the data on a chart of the semirange of the cyclic stress component as the ordinate and the mean stress as the abscissa, is called the s_a–s_m diagram. Various failure criteria are defined on this diagram as curves joining the two axes drawn from the (s_m, s_a) point $(0, s_a)$ to $(s_Y, 0)$ or to $(s_u, 0)$ depending on the criterion chosen, where sY is the apparent yield (or proof) stress and su is the ultimate stress. For anisotropic materials the stress direction in the material is important. These curves seek to define, for a given endurance limit, the boundary within which the stress combination is deemed to be safe and outside of which the combination will be unsafe. The three criteria widely quoted in standard texts are termed the Gerber, the modified Goodman and the Soderberg criteria. They are deterministic in nature and in practice are associated with assumed factors of safety whose magnitude is intuitive and depends in an arbitrary way upon the perceived value at risk.

The solution to a problem in these terms is an exact number of cycles that is sufficiently distant from the failure boundary to provide a satisfactory level of confidence for the design management.

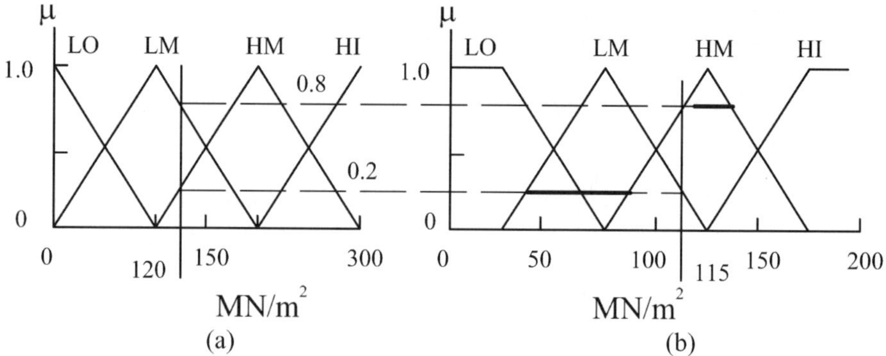

Figure 6.3. Partitioning of the stress UDs. (a) Mean stress s_m. (b) Alternating stress s_a.

6.2. Fuzzy Analysis

6.2.1. COMBINED STRESSES

The experimental data associated with combined stress fatigue normally exhibits significant scatter which makes the decision of the location and shape of the failure criterion as a distinct curve on the (sm,sa) diagram, difficult. The actual stresses at a point in a design may also be difficult to determine. In the FL treatment what emerges is a broad view in which the fine detail with a high level of uncertainty is absent. The uncertainty in the knowledge of the physical processes and of the input data in practice is carried through the logic processes and remains explicit in the conclusions.

The method in the FL treatment is to partition the (s_m, s_a) UDs into fuzzy sets. The coarseness of the partitioning reflects the level of uncertainty in the experimental data; the less the uncertainty the more fuzzy sets can be defined on each UD. The heuristic array of rules which is formed relates the fuzzy sets on the two UDs, they are synthesised from a palette of available diagnostic physical data, idealised models and analysis, human knowledge and experience of similar physical systems. An example of possible partitioning of the (s_m, s_a) UDs for an aluminium is shown in Figure 6.3.

The membership functions corresponding with Figure 6.3(a) are

	$0 \le s_m \le 100$	$100 \le s_m \le 200$	$200 \le s_m \le 300$
μ_{LO}	$1 - s_m/100$		
μ_{LM}		$s_m/100$	$2 - s_m/100$
μ_{HM}		$(s_m - 100)/100$	$3 - s_m/100$
μ_{HI}			$(s_m - 200)/100$

Table 6.2. A combined stress relational array.

s_a	LO	LM	HM	HI
s_m	HI	HM	LM	LO

(LO = low; LM = low medium;
HM = high medium; HI = high)

The membership functions corresponding with Figure 6.3(b) are

	$0 \leq s_a \leq 25$	$25 \leq s_a \leq 75$	$75 \leq s_a \leq 125$	$125 \leq s_a \leq 175$
μ_{LO}	1.0	$1.5 - s_a/50$		
μ_{LM}		$(s_a - 25)/50$	$2.5 - s_a/50$	
μ_{HM}		$(s_a - 75)/50$	$3.5 - s_a/50$	
μ_{HI}			$(s_a - 125)/50$	

The type of rule array appropriate for combined stress fatigue is shown in Table 6.2.

EXAMPLE 6.1. Consider a stepped shaft which has been fine machined from aluminium stock bar, as shown in Figure Ex 6.1.

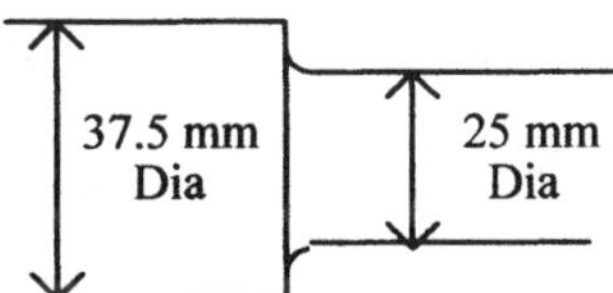

Figure Ex 6.1. Aluminium stepped shaft.

The shaft is subject to fatigue failure in rotation about its axis due to a uniform bending moment in the plane of the paper and also a steady axial tensile force of 36.78 kN. The elastic stress concentration factor for the fillet is assumed to be $K_t = 1.6$ and the fatigue reduction factor is assumed to be $K_f = 1.5$. Diagnostic laboratory fatigue tests on specimens of the same material machined to a similar finish and of the same scale give an average endurance limit (10^6 cycles) of 200 MN/m^2 and an average linear elastic limit of 300 MN/m^2.

Find the susceptibility of the shaft to fatigue failure due to the bending moment.

Solution. It will be assumed that the partitioning of the (s_m, s_a) UDs shown in Figure 6.3 is valid, also that the relationship array in Table 6.2 is appropriate.

From basic solid mechanics theory the nominal tensile stress in the smallest diameter is 74.88 MN/m^2 and applying the stress concentration factor gives a stress peak of 120 MN/m^2. This value is shown in Figure 6.3(a). The fuzzified form is 0.8LM, 0.2HM. The FL proposition is

80   *Fuzzy Logic Applications in Engineering Science*

IF S_m	THEN S_a	MEM. VALUE	CONSEQUENCE
LM	HM	0.8	0.8HM
HM	LM	0.2	0.2LM

The conclusion is the union of the consequences

$$S = 0.2\text{LM} \cup 0.8\text{HM}.$$

This is displayed in Figure 6.3(b). It may be noted that the major contribution is in the HI set with a residual effect in the LM set. This means that there is possibly some risk of fatigue failure from a low alternating stress of 25 MN/m^2 (16.75 MN/m^2 nominal stress). Beyond an alternating stress of 175 MN/m^2 (117 MN/m^2 nominal stress) it is unlikely that any of these components would survive. These stress values correspond with bending moments of 38 and 267 kN m respectively. The EFN has a value of 115 MN/m^2 and is shown in Figure 6.3(b). Defuzzifying the conclusion would give a similar value.

For comparison, the conventional diagram with a Soderberg criterion gives an alternating stress value of 120 MN/m^2 (80.2 MN/m^2 nominal stress). The nominal stress would normally be used with a factor of safety. If it was desired that no failures would be likely, then the factor of safety would need to be $120/25 = 4.8$. Usually, the factor of safety is chosen arbitrarily. The susceptibility of the design to fatigue is shown in Figure 6.3(b). The design decision depends upon what level of risk is acceptable.

6.2.2. ENDURANCE LIMITS

The fatigue endurance limit for a particular material at a given stress level is normally given as a precise value. But it will be recognised that this is idealistic and at best representing a tendency rather than a definite fact. By casting the discussion in FL terms the full scope of the possibilities emerges.

For the aluminium considered above, the diagnostic s_a-log n fatigue data shows a bilinear form. Figures 6.4(a) and (b) illustrate the partitioning of the UDs. The corresponding heuristic relationship array is shown in Table 6.3.

The membership functions for Figure 6.4(a) are

	$0 \le s_a \le 90$	$90 \le s_a \le 100$	$100 \le s_a \le 120$	$120 \le s_a \le 150$	$150 \le s_a \le 250$
μ_{LO}	1.0	$10 - s_a/10$			
μ_{LM}		$(s_a - 90)/10$	$6 - s_a/20$		
μ_{HM}			$(s_a - 100)/20$	$5 - s_a/30$	
μ_{HI}				$(s_a - 120)/30$	$2.5 - s_a/100$
μ_{VH}					$(s_a - 150)/100$

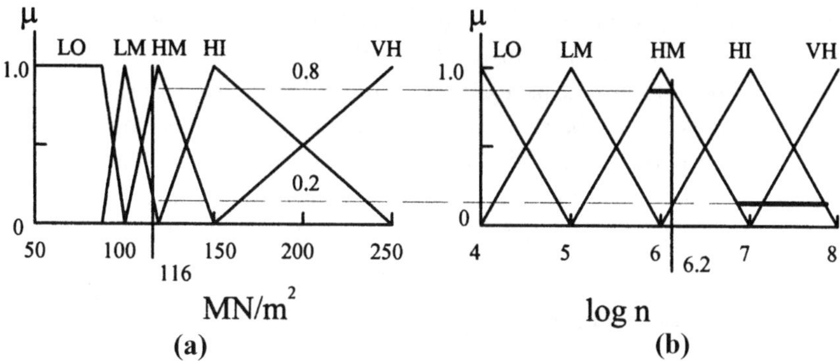

Figure 6.4. Partitioning of the UDs. (a) Alternating stress. (b) Log number of cycles.

Table 6.3. Endurance limit relational array.

s_a	LO	LM	HM	HI	VH
log n	VH	HI	HM	LM	LO

(LO = low; LM = low medium; HM = high
medium; HI = high; VH = very high)

The membership functions for Figure 6.4(b) are

	$4 \leq \log n \leq 5$	$5 \leq \log n \leq 6$	$6 \leq \log n \leq 7$	$7 \leq \log n \leq 8$	$8 \leq \log n \leq 9$
μ_{LO}	$5 - \log n$				
μ_{LM}	$\log n - 4$	$6 - \log n$			
μ_{HM}		$\log n - 5$	$7 - \log n$		
μ_{HI}			$\log n - 6$	$8 - \log n$	
μ_{VH}				$\log n - 7$	$9 - \log n$

EXAMPLE 6.2. In the previous example it was found that the EFN of the alternating stress at the endurance limit of 106 cycles was 116 MN/m². Find the number of cycles corresponding with this alternating stress.

Solution. Assuming the partitioning of the alternating stress and log number of cycles as shown in Figures 6.4(a) and (b) respectively. Also the relational array in Table 6.3.

From Figure 6.4(a) the membership values are: $\mu_{LM} = 0.2$ and $\mu_{HM} = 0.8$. The FL proposition is

IF S_a	THEN LOG N	MEM VALUE	CONSEQUENCE
LM	HI	0.2	0.2HI
HM	HM	0.8	0.8HM

The conclusion is, $\log n = 0.2\text{HI} \cup 0.8\text{HM}$.

This conclusion is illustrated in Figure 6.4(b). It may be noted that the major portion of the conclusion lies in the HM set of $\log m$ which is centred about 6 on the UD. There is also a minor part in the HI set. The EFN is 6.2, which is $1.58 * 10^6$ cycles, this would need to be compared with the value at risk from the maintenance perspective. Thus there is a tendency for the shaft to exceed the required endurance limit, but there is the possibility of failure from 10^5 cycles. There is a small possibility that the component may survive to about 10^8 cycles.

6.3. Low-Endurance Fatigue

In low endurance fatigue the material is cyclically strained beyond that elastic limit. This leads to a lower endurance limit, perhaps only of the order of several thousand cycles or less. For systems that are only cycled between a few or in the low hundreds per annum, this means that loading in the plastic range can still provide a useful and economic service life. Such cases are more commonly found in larger structures and rarely found in consumer items.

Diagnostic investigations are often reported in terms of plastic strain rather than stress. Surface strains are directly measurable, but stresses would have to be inferred using an assumed rheological (or constitutive) equation, which is less reliable than in the case of small linear elastic strains associated with high-endurance fatigue. An empirical deterministic fatigue correlation has been found to describe the trend of many metals, it is of the form

$$\varepsilon_p n^q = b. \tag{6.2}$$

This is analogous to the high-endurance fatigue expression, Equation (6.1). The pattern of the FL treatment described in Section 6.2 for high-endurance fatigue may provide methodology in some cases for low-endurance fatigue. This has not been developed up to the present time, but clearly offers the same advantages in providing richer conclusions to problems compared with the conventional treatment.

6.4. Creep

Laboratory diagnostic data for creep are usually correlated with an empirical expression comprising strain as the product of functions of stress (s), time (t) and temperature (θ)

$$\varepsilon = f_1(s) f_2(t) f_3(\theta). \tag{6.8}$$

More complex formulae with some theoretical basis and with sufficient disposable constants for curve fitting have also been suggested.

For metals, the experimental data is grouped into:

(i) an initial elastic deformation, which is relatively small,

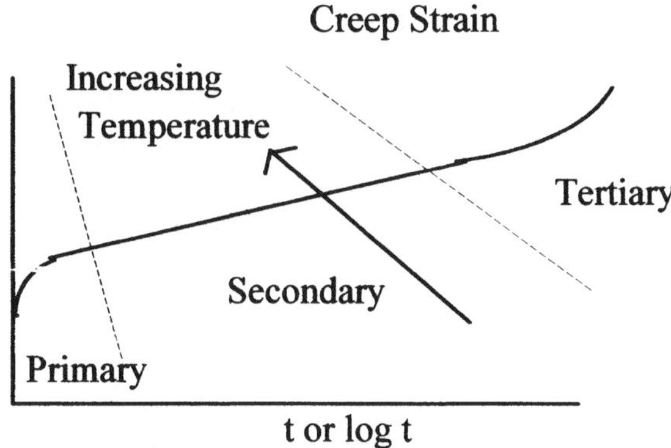

Figure 6.5. Stages in the creep of metals.

(ii) a primary phase of decreasing plastic strain rate,
(iii) a secondary phase of steady plastic strain rate, and
(iv) a tertiary phase with an accelerating plastic strain rate until rupture occurs.

These phases are illustrated in Figure 6.5. Engineering design beyond the secondary phase is not normally undertaken.

Experimental data for the secondary stage of creep often appear to exhibit a near linear relationship with time or log time, depending upon the case. A relationship between the creep strain and time spanning the primary and secondary stages may be expressed as

$$\varepsilon - \varepsilon_1 = \mu^a (km + \varepsilon_0), \tag{6.9}$$

where ε_1 is the strain at the onset of inelastic flow, k is the slope of the secondary stage of the graph, m is t or $\log t$ depending upon the case and ε_0 is the intercept of the secondary stage line with the vertical axis. The membership function (μ) is defined by

$$\mu = (\varepsilon - \varepsilon_1)/(\varepsilon_2 - \varepsilon_1) \quad (\varepsilon_1 \leq \varepsilon \leq \varepsilon_2). \tag{6.10}$$

Strain ε_2 defines the end of the primary stage. In Equation (6.9) the index a is determined from a fair point in the primary stage data.

Consonant with the form of Equation (6.8), various empirical deterministic stress-strain relations have been proposed for the primary stage of deformation such as

$$s = c\varepsilon_p^q. \tag{6.11}$$

But the alternative treatment above based upon a membership function is more flexible and can apply over a more extensive range of variables. However, it is still deterministic in nature. The temperature factor is again often described by an empirical series type of expression with sufficient disposable constants to facilitate curve fitting of the experimental data.

The relationship implied in Equation (6.8) may be expressed in FL terms, thus avoiding the need for precise mathematical expressions for the functions $f_1(s)$, $f_2(t)$ and $f_3(\theta)$. In the FL framework, the relationship would be expressed in the form of a two-dimensional array with set parameters of temperature and time (or logtime). The corresponding FL proposition is

$$\text{IF } \theta \quad \text{AND T} \quad \text{THEN E} \quad \text{MIN} \quad \text{CONSEQUENCE}$$

As in all other cases, the FL conclusion yields knowledge of the range of risks in a given design with the selected material.

6.5. Remarks

Even with carefully prepared specimens it is found that there is significant scatter in the experimental results of fatigue tests of engineering materials. Assuming a mean line correlation to provide an empirical deterministic relation inevitably discards useful and significant information. A statistical analysis is based upon an assumed probability distribution (assuming that a sufficient number of sample results are available), which may or may not be appropriate. No formulae are assumed in the FL treatment and it is therefore more general than the alternatives. The FL framework is a synthesis of the knowledge of trends, accuracy of data and knowledge of similar systems. Uncertainty or vagueness in the information is carried through to the conclusions and therefore engineering design management is able to make decisions based upon knowledge of the risks. Similar remarks apply to the creep phenomenon.

In the case of large one-of-a-kind designs, it may not be possible to test the whole artefact because of scale and temporal considerations. Then resort may be made to model tests. There would generally be too few results for statistical analysis, but a FL approach would be feasible. In any case it is necessary to be aware of scale effects, whereby a large scale structure may exhibit different properties to small scale test pieces.

Another reason in favour of the FL method is that the service conditions are often not know with any precision. The FL method is more adaptable to this factor.

The more fundamental approach of fracture mechanics offers no advantages because the formulae modelling the processes assume an idealised continuum which is not like the micro structure of a real engineering material. Also, because this treatment depends upon other assumptions, including the size of the initial flaw in the material and the fatigue crack shape, it is not recommended in British Standard BS 7608 that it be used to define fatigue strengths or lives.

Chapter 7
Cumulative Fatigue Damage Analysis

It is infrequently found that a component or structure is only subjected to a uniform sinusoidal load pattern throughout its service life. It is much more usual for the load to be of variable amplitude in a pattern which may or may not repeat itself. If there is simple repition it may be possible in some cases to emulate this in laboratory specimen material or in model or prototype diagnostic tests. This can be time consuming and it may be impractical to plan such tests, which with some structures could take many years. Another factor is that it may be difficult to replicate the service environmental conditions for models and prototypes, which may include interacting causal factors such as inertia forces, vibrations, pressure changes, thermal effects, creep, corrosion, ground movement and also freeze/thawing and moisture fluctuations in concrete structures.

At the design stage it is necessary to predict the likely fatigue behaviour of an artefact and it would be useful if this could be accomplished by the application of information from specimen material fatigue laboratory tests, which are frequently obtained from high-endurance fatigue tests under a constant amplitude load. This may be possible given the following information:

(i)   The stress fields at the critical points within the design.
(ii) The microstructure at the critical points.
(iii) Laboratory specimen material fatigue test data for similar stress fields with a similar microstructure.
(iv) A criterion for aggregating the effects of varying load patterns.
(v)   A method of counting the number of cycles in a variable loading pattern.

In considering items (i) and (iii) above it may be noted that for nominally identical test specimens

(a) The fatigue strength is usually greatest in bending, followed by axial tension then torsion.
(b) The fatigue limit follows the same order as above.

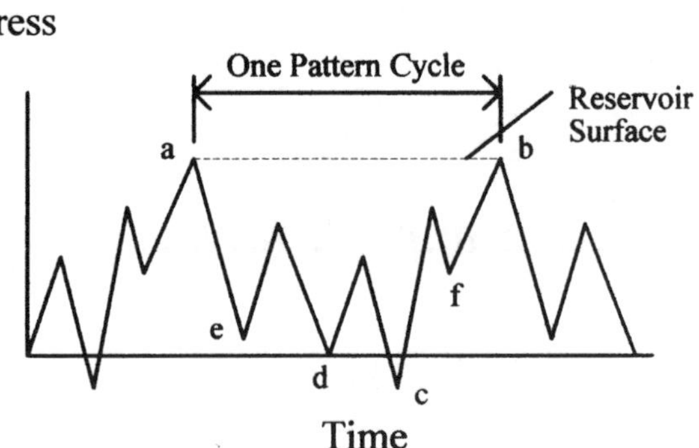

Figure 7.1. A repetitive stress pattern.

It may be further noted that in wrought metals the fatigue resistance (and static strength) across the grain may be about 85% of that along the grain. Also, that a coarse grained microstructure offers a higher resistance to fatigue than a fine grained one of the same material.

Various methods have been suggested for aggregating the effects of a variable load pattern, the most widely quoted is Miner's hypothesis, which is treated in some detail in this chapter. Two methods are used in practice to count the number of cycles in a mixed load pattern, these are known as the Rainflow and the Reservoir methods. They both give similar results in the long term and only the Reservoir method is described here.

7.1. Cycle Counting

In most cases, some or all of the inservice loading history will have a repetitive content. If this can be identified then it is possible to count the number of cycles for each of a series of load amplitude levels by applying the Reservoir method, which is advocated by British Standard BS5400: Part 10: 1980 (for example). There is also computer software available that will provide similar information.

In the Reservoir method it is necessary to have a real or simulated record of the load history (or stress at a critical point in a design) over a sufficiently long period of time to establish a stable acceptable average pattern, such as that illustrated in Figure 7.1.

The method is as follows:

(i) To assist identification, the high points, (a, b) in adjacent stress patterns may be joined by a line, as shown. This represents the surface of an imaginary full reservoir.

(ii) Drain the water from the lowest point c. The distance fallen by the water surface line a–b represents the first stress level, s_1.

(iii) Drain the water from the next lowest point, d. The distance fallen by the water surface is the second stress level, s_2.

(iv) Continue draining from successively lowest points until no water is remaining, to find the remaining stress levels s_3 and s_4, etc.

It is then possible to compile a list of the number of stress cycles at each stress level per pattern cycle, and by this means to find the number of stress cycles at each stress level over a given service period.

7.2. Miner's Rule

The most widely quoted hypothesis for fatigue damage aggregation to date is known as Miner's rule (also known as the Palmgren–Miner rule). None of the alternatives have achieved the same level of acceptance or the status of being incorporated into Standards such as BS 5400: Part 10 and BS 7608, although it is recognised that in practice the rule is not unequivocally successful. But the simplicity of the rule is one of its attractions and variants of the rule appear to show no clear advantages. Miner's rule is normally expressed in the form

$$\sum n_i / n_{ip} = 1, \tag{7.1}$$

where $1 \leq i \leq k$. n_i is the number of repetitions of the applied stress range, s_i and n_{ip} is the number of repetitions of the same applied stress range to failure. This formula is normally associated with the empirical s–n fatigue correlation discussed in Chapter 6. In Equation (7.1) the n_i are not constrained to any particular sequence, nor is each of the ni necessarily in monobloc form according to BS 5400: Part 10 and BS 7608 and any of the other available publications. It is found in practice that the summation in Equation (7.1) may be more or less than unity by as much as 20% or more. There is, of course, significant scatter in the experimental results and n_{ip} is not easy to determine. The stress range content, or wave form, is not specified, although in practice, much of the laboratory test data is gathered under sinusoidal load (or deformation) conditions.

Both the above Standards also identify a threshold stress range, s_o, which is the stress range at a specific endurance and is considered to be the boundary of the "non-propagating stress range" below which the material can sustain an indefinitely large number of cycles in clean air (or more generally, a passive environment). It is noted this does in fact depend upon any modifying factors required to the $(s$–$n)$ relationship due to material thickness, the effect of the environment on any unprotected joints and the effect of weld grinding in steel structures. For example, for unprotected joints in sea water the recommendation is that $s_o = 0$, that is, there is no fatigue limit. For $0 < s < s_o$ the index of the s–n relation is modified to $(m + 2)$.

In the case where the applied fluctuating load is of varying magnitude such that some of the $s_i < s_o$ will cause crack propagation, resulting in an earlier fatigue failure than if all $s_i < s_o$ were non-contributory. Both Standards advocate the use of a weight factor $(s_i/s_o)^2$ applied to the appropriate cycle ratio. Thus the effect of $(s_i/s_o)^2$ becomes vanishingly small as s_i/s_o approaches zero.

The Standards also specify that for acceptability, the summation in Equation (7.1) should not exceed 1.0 at any design point. In the case where the limit is exceeded it is recommended that the artefact design should be amended by strengthening or redesigning until new stress ranges is found that are less than the original stresses divided by $\alpha^{1/m}$ and $\alpha^{1/(m+2)}$, where α represents the summation in Equation (7.1).

BS 7910 for fusion welded structures is another Standard which refers to Miner's rule. It specifies $2*10^6$ cycles on the standard $(s\text{--}n)$ correlation at constant stress range (s_r) as a reference for mixed stress ranges. It also defines an equivalent stress (s_e) for mixed stress ranges

$$s_e = \left\{ \left(\sum s_i^3/n_{ip} \right) \Big/ 2*10^6 \right\}^{1/3}. \tag{7.2}$$

7.3. Fuzzy Logic Treatment

7.3.1. METHODOLOGY

In the following discussion, the fatigue damage aggregation is cast in terms of fuzzy sets rather than in terms of a deterministic formula, as in Miner's rule. It is generally accepted that fluctuating loads increase the proclivity of a metal to fail and that the metal is increasingly "damaged" by the process. Furthermore, it is reported that if a metal is subjected to a higher fluctuating load range, followed by a lower one, then the effect of the earlier higher loading is to accelerate the damage due to the lower load range.

In this work, the area on the $s\text{--}n$ chart containing the failure data is called the p-zone and it represents the zone of limiting material damage. Similarly, there is a nucleation/crack growth transition zone, which represents the material state of incipient damage, called here the m-zone. These two zones are illustrated in Figure 7.2. Between these two zones is a region of progressive material damage as the number of stress cycles increases.

Also shown in Figure 7.2 is the upper stress limit, s_u, which represents the elastic stress limit and also the threshold stress, s_o, which is the upper boundary of non-propagating stress range.

In the FL portrayal of fatigue, the stress ranges in the m and p-zones are represented as fuzzy sets and related to the corresponding number of cycles, which are also represented as fuzzy sets. The relationship is expressed as a heuristic relational array, such as that shown in Table 7.1 and a logic proposition of the form

IF S   THEN N

The fuzzy sets are defined by partitioning of the stress and number of cycles UDs, the latter according to whether it refers to the m or p-zone, as shown below. The

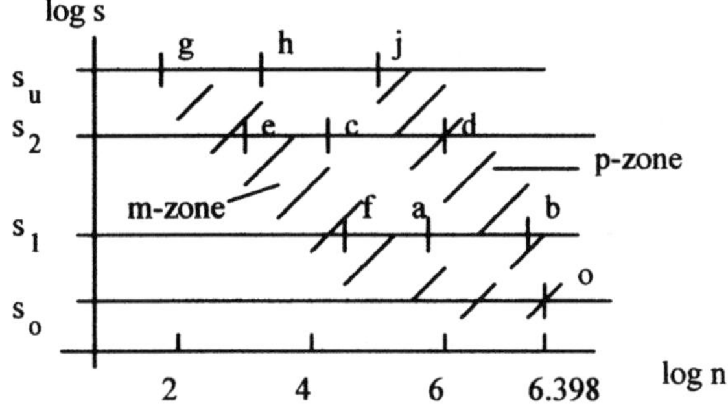

Figure 7.2. Illustration of the material damage zones.

Table 7.1. Relational array for high-endurance fatigue.

No. Cycles Sets	Stress Sets				
	S_1	S_2	S_3	S_4	S_5
M-zone	M1	M2	M3	M4	M5
P-zone	P1	P2	P3	P4	P5

number of fuzzy sets defined on each UD depends upon the degree of precision of the information available, the greater the precision, the finer the partitioning possible.

A more general treatment is achieved by defining scaled variables as follows (see Figure 7.2 for reference points). The stress metric element (s_i^+) is defined by

$$s_i^+ = \log(s_u/s_i^+)/\log(s_u/s_o). \tag{7.3}$$

The number of cycles metric element is defined by

$$n_p^+ = \log(n_p/n_{jp})/\log(n_o/n_j) \tag{7.4}$$

for the maximum damage *p*-zone or by

$$n_m^+ = \log(n_m/n_j)/\log(n_o/n_j) \tag{7.5}$$

for the incipient damage *m*-zone.

Within the damaged region on the $\log n$–$\log s$ chart, a material damage metric is defined in terms of the fractional number of cycles by

$$d^+ = \log(n/n_m)/\log(n_p/n_{mp}), \tag{7.6}$$

where n_m and n_p are characteristic numbers of cycles in the *m* and *p*-zones respectively.

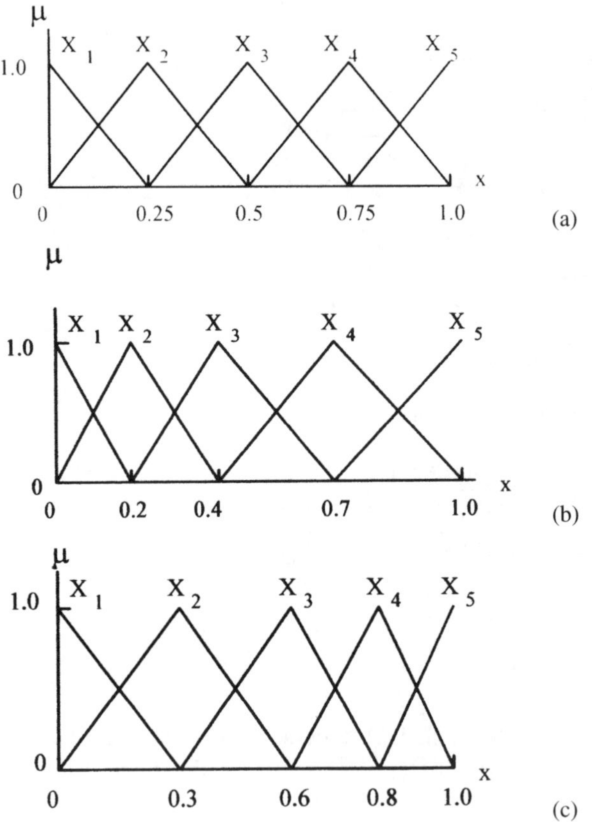

Figure 7.3. Typical five set partitioning patterns. (a) Equipartitioned. (b) Left-Biased. (c) Right-Biased.

Examples of uniform (equipartitioned) and non-uniform partitioning of the general UD (x) are illustrated in Figures 7.3(a), (b) and (c) for five fuzzy sets in each case.

The partitioning patterns shown in Figures 7.3(a), (b) and (c) have membership functions given below.

Membership functions for Figure 7.3(a)

	$0 \leq x \leq 0.25$	$0.25 \leq x \leq 0.5$	$0.5 \leq x \leq 0.75$	$0.75 \leq x \leq 1.0$
μ_1	$1 - x/0.25$			
μ_2	$x/0.25$	$2 - x/0.25$		
μ_3		$(x - 0.25)/0.25$	$3 - x/0.25$	
μ_4			$(x - 0.5)/0.25$	$4 - x/0.25$
μ_5				$(x - 0.75)/0.25$

Membership functions for Figure 7.3(b)

	$0 \le x \le 0.2$	$0.2 \le x \le 0.4$	$0.4 \le x \le 0.7$	$\le 0.7 \le x \le 1.0$
μ_1	$1 - x/0.2$			
μ_2	$x/0.2$	$2 - x/0.2$		
μ_3		$(x - 0.2)/0.2$	$(0.7 - x)/0.3$	
μ_4			$(x - 0.4)/0.3$	$(1 - x)/0.3$
μ_5				$(x - 0.7)/0.3$

Membership functions for Figure 7.3(c)

	$0 \le x \le 0.3$	$0.3 \le x \le 0.6$	$0.6 \le x \le 0.8$	$0.8 \le x \le 1.0$
μ_1	$1 - x/0.3$			
μ_2	$x/0.3$	$2 - x/0.3$		
μ_3		$(x - 0.3)/0.3$	$4 - x/0.2$	
μ_4			$(x - 0.6)/0.2$	$5 - x/0.2$
μ_5				$(x - 0.8)/0.2$

7.3.2. APPLICATIONS

The focus in this discussion is on the aggregation of fatigue damage due to mixed load ranges and as a primary example of this a step change in the load range is considered, both for increasing and decreasing loads at several damage levels. The case of a structural steel will be considered and referring to Figure 7.2, typical fatigue values would be given by

$$s_u = 630 \text{ MPa}, \quad n_j = 104 \text{ cycles},$$

$$s_o = 100 \text{ MPa}, \quad n_o = 2.5 * 10^6 \text{ cycles}.$$

There is very little information about the onset of incipient fatigue damage at different stress levels. Therefore, it will be assumed that it ranges from about 10^2 cycles at the high stress level to $2.5 * 10^6$ cycles at the non-propagating stress level (s_o).

Salient points (g, j, o) on the $\log s$–$\log n$ chart are identified in Figure 7.2. In this diagram, s_1 and s_2 are two successively applied stress levels in alternate order. Points a, c and h refer to a common damage state at the stress levels s_1, s_2 and s_u respectively. Points b, d, e and f are found below in the FL treatment.

Case 1. Left-biased partitioning. Salient points.
Salient points in Figure 7.3 are initially found using the left-biased partitioning in Figure 7.2(b). In the first application, the initial stress range is $s_2 = 222$ MPa, with a step change in stress level to 132 MPa at several damage states d_a^+, defined by Equation (7.6), in the range $d_f^+ \le d_a^+ \le d_b^+$.

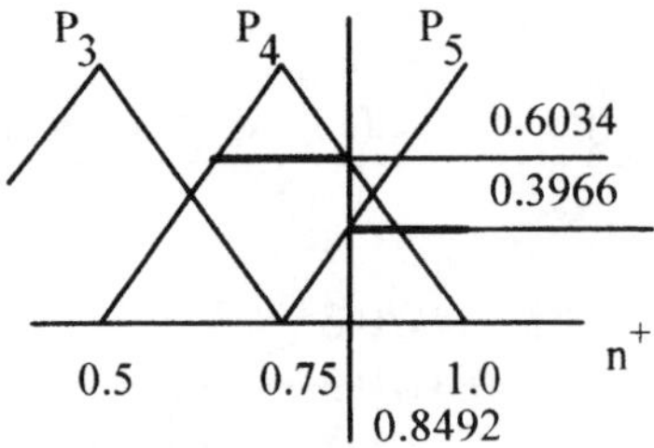

Figure 7.4. p-zone conclusion.

From the stress range values given and Equation (7.3), the following stress metric values may be found:

$$s_o^+ = 1.0; \quad s_1^+ = 0.8492; \quad s_2^+ = 0.5667; \quad s_u^+ = 0.0.$$

Now assuming that the stress metric UD is equipartitioned into five fuzzy sets, as illustrated in Figure 7.3(a). The membership values for the stress metrics above may be found from the membership functions. Their values are as follows:

$$\text{For } s_o^+ \quad \mu_{s5} = 1.0$$
$$s_1^+ \quad \mu_{s4} = 0.6034; \mu_{s5} = 0.3966$$
$$s_2^+ \quad \mu_{s3} = 0.7332; \mu_{s4} = 0.2668$$
$$s_u^+ \quad \mu_{s1} = 1.0$$

For the *p*-zone the form of the FL proposition is

IF S	THEN P	MEM. VALUE	CONSEQUENCE
S_4	P_4	0.6034	$0.6034P_4$
S_5	P_5	0.3966	$0.3966P_5$

The conclusion is the union of the consequences, thus,

$$P = 0.6034P_4 \cup 0.3966P_5. \tag{7.7}$$

The interpretation of this expression requires an explicit partitioning of the endurance UD. For this purpose the pattern shown in Figure 7.3(a) is again chosen. The result is shown in Figure 7.4.

The major contribution is in the P_4 fuzzy set, with a lesser contribution in the P_5 fuzzy set. There is a tendency for the value of n^+ to lie around the EFN of 0.8492, which is a scaled value of the number of cycles in the *p*-zone corresponding with a stress range of 132 MPa. But the range of the conclusion illustrated in Figure 7.4 must be remembered in design work. The physical number of cycles is obtained by rearranging Equation (7.5):

$$n_b = n_j(n_o/n_j)^{0.8492} = 10^4(2.5 * 10^6/10^4)^{0.84926}.$$

Hence,

$$n_b = 1.0870 * 10^6 \text{ cycles}.$$

Similarly, the FL proposition for s_2^+ is

IF S	THEN P	MEM. VALUE	CONSEQUENCE
S_3	P_3	0.7332	$0.7332P_3$
S_4	P_4	0.2668	$0.2668P_4$

As before, the conclusion is the union of the consequences

$$P = 0.7332P_3 \cup 0.2668P_4. \tag{7.8}$$

The EFN is 0.5667 and the physical number of cycles corresponding with this dimensionless value is $2.2851 * 10^6$ cycles.

Values of the number of cycles in the m-zone may also be found for the stress metrics s_1^+ and s_2^+. Since the nature of the m-zone is less clear than for the p-zone, two partitioning patterns will be considered, namely the left-bias and the right-bias as illustrated in Figures 7.3(b) and 7.3(c) respectively. In the present case the left-bias pattern will be used, the right-bias pattern will be used in Case 2 below for comparison.

The same procedure outlined below is used for the m-zone and the p-zone. For s_1^+

IF S	THEN M	MEM. VALUE	CONSEQUENCE
S_4	M_4	0.6034	$0.6034M_4$
S_5	M_5	0.3966	$0.3966M_5$

The conclusion is

$$M = 0.6034M_4 \cup 0.3966M_5. \tag{7.9}$$

The EFN is $n_f^+ = 0.8190$, which yields a physical value

$$n_f = n_g(n_o/n_g)^{0.8190} = 10^2(2.5 * 10^6/10^2)^{0.8190}$$

$$= 3.998 * 10^5.$$

Also, for s_2^+

IF S	THEN M	MEM. VALUE	CONSEQUENCE
S_3	M_3	0.7332	$0.7332M_3$
S_4	M_4	0.2668	$0.2668M_4$

The conclusion is

$$M = 0.7332M_3 \cup 0.2668M_4. \tag{7.10}$$

Table 7.2. Summary of Case 1 characteristic no. of cycles.

No. Cycles	Stress Range MPa			
	100	132	222	630
p-zone	$2.5 * 10^6$	$1.087 * 10^6$	$2.285 * 10^5$	10^4
m-zone	$2.5 * 10^6$	$3.998 * 10^5$	$1.292 * 10^4$	10^2

The EFN is $n_e^+ = 0.4800$. The physical corresponding physical value is

$$n_e = n_g(n_o/n_g)^{0.4800} = 10^2(2.5 * 10^6/10^2)^{0.4800}$$

$$= 1.2918 * 10^4.$$

The above results are summarised in Table 7.2.

The values of n_b, n_d, n_e and n_f found above are tendencies not deterministic precise values. The fuzzy sub-set conclusions represent the full solutions.

It may be recalled that the m-zone and p-zone represent the lower (effectively zero) and upper bounds of material damage states. To investigate the effect of mixed loading patterns consideration is now given to step changes in the stress ranges at different damage states, where the damage is defined by the metric d^+.

Case 1. Mixed loading patterns.
Consider a material damage level, $d^+ = 0.2$, then if the points a, c and h in Figure 7.3 represent states of equal material damage, then from Equation (7.6)

$$d_h^+ = \log(n_h/n_g)/\log(n_j/n_g), \tag{7.11}$$

$$d_c^+ = \log(n_c/n_e)/\log(n_d/n_e) \tag{7.12}$$

and

$$d_a^+ = \log(n_a/n_f)/\log(n_b/n_f). \tag{7.13}$$

Rearranging Equations (7.11)–(7.13) and inserting numerical values yields

$$n_h = 10^2(10^4/10^2)^{0.2} = 2.512 * 10^2,$$

$$n_c = 1.292 * 10^4(2.285/0.1292)^{0.2} = 2.295 * 10^4,$$

$$n_a = 3.998 * 10^5(1.087/0.3400)^{0.2} = 4.884 * 10^5.$$

Now it may easily be shown that if the sum of the fractional cycles increment is less for a step decrease than for a step increase then $n_a/n_b > n_c/n_h > n_h/n_j$. For example, for a step change between s_1 and s_2, if Path $ecab$ < Path $facd$ (Figure 7.3), then

$$n_c/n_d + (n_b - n_a)/n_b < n_a/n_b + (n_d - n_c)/n_d. \tag{7.14}$$

Table 7.3. Fractional cycles summation values for Case 1.

Damage Metric d_a^+	Load Path			
	ecab	*facd*	*echj*	*ghcd*
0	0.6887	1.311	1.047	0.9535
0.2	0.6512	1.349	1.075	0.9247
0.4	0.6296	1.370	1.115	0.8847
0.65	0.6612	1.339	1.166	0.8337
0.8	0.7443	1.256	1.165	0.8352
0.9	0.8455	1.155	1.119	0.8807
1.0	1.0	1.0	1.0	1.0

Hence, $n_a/n_b > n_c/n_d$, etc.

Therefore, the rule is

$$n_a/n_b > n_c/n_d > n_h/n_j. \tag{7.15}$$

Finding the fractional number of cycles summation to failure for step changes of stress range between s_1 and s_2 and between s_2 and s_u at a damage level, $d_a = 0.2$ yields the following values:

$$n_a/n_b = 0.4493; \quad n_c/n_d = 0.1004; \quad n_h/n_j = 0.02512.$$

Thus the inequality, expression (7.14) is satisfied. Now calculating the several path lengths:

Path

$$ecab \quad 1 + n_c/n_d - n_a/n_b = 1.100 - 0.4493 = 0.6512, \tag{7.16}$$

$$facd \quad 1 + n_a/n_b - n_c/n_d = 1.449 - 0.1004 = 1.349, \tag{7.17}$$

$$echj \quad 1 + n_c/n_d - n_h/n_j = 1.100 - 0.0251 = 1.075, \tag{7.18}$$

$$ghcd \quad 1 + n_h/n_j - n_c/n_d = 1.025 - 0.1004 = 0.9247. \tag{7.19}$$

These results agree with the inequality expression (7.14).

A similar analysis may be conducted for other values of the damage metric d_a^+. Results for increasing d_a^+ values of 0, 0.2, 0.4, 0.65, 0.8, 0.9 and 1.0 are given in Table 7.3.

It may be noted in Table 7.3 that for all the above damage metrics a step reduction in stress range corresponds with a fractional cycles summation < 1.0, whereas for a step increase the summation is > 1.0. Furthermore, for adjacent pairs of columns the average value is 1.0.

Case 2. Right-biased partitioning. Salient points.

It is not clear at this point whether the relationship between the step changes in stress range and the values listed in Table 7.3 are governed by the choice of the left-biased partitioning of the number of cycles UD in the m-zone. As a check an alternative right-biased partitioning, as illustrated in Figure 7.2(c) has been used in the following parallel investigation.

The membership functions of the right-biased partitioning have been given before. New values of n_e, n_f and n_u are found below. Other values remain as before. The fuzzy sets for s_1^+ and s_2^+ are as before, namely,

$$\text{For } s_1^+: \quad M = 0.6034M_4 \cup 0.3966M_5.$$
$$\text{For } s_2^+: \quad M = 0.7332M_3 \cup 0.2668M_4.$$

Using the EFN as a tendency value for the number of cycles, it may be shown that these are

$$n_f^+ = 0.8793 \quad \text{and} \quad n_e^+ = 0.6534.$$

The physical values of n_e and n_f are given by

$$n_e = n_g(n_o/n_g)^{0.6534} = 10^2(2.5 * 10^6/10^2)^{0.6534}$$
$$= 7.366 * 10^4 \text{ cycles}$$

and

$$n_f = n_g(n_o/n_g)^{0.8793} = 10^2(2.5 * 10^6/10^2)^{0.8793}$$
$$= 7.472 * 10^5 \text{ cycles}.$$

As before, $n_b = 1.087 * 10^6$ cycles and $n_d = 2.285 * 10^5$ cycles.

Now considering again a damage metric, $d_a^+ = 0.2$. Rearranging Equations (7.11)–(7.13) and inserting numerical values yields

$$n_h = 10^2(10^4/10^2)^{0.2} = 2.512 * 10^2 \quad \text{(as before)},$$

$$n_c = 7.472 * 10^4(2.285/0.7472)^{0.2} = 9.344 * 10^4,$$

$$n_a = 7.366 * 10^5(1.087/0.7366)^{0.2} = 7.962 * 10^5.$$

Checking the inequalities (7.14)

$$n_a/n_b = 0.7325; \quad n_c/n_d = 0.4089; \quad n_h/n_j = 0.02512.$$

The fractional number of cycles summation for stress range steps between s_1 and s_2 and between s_2 and s_u for the above damage level yields

Path

$$ecab \quad 1 + n_c/n_d - n_a/n_h = 1.409 - 0.7325 = 0.6764, \tag{7.20}$$

$$facd \quad 1 + n_a/n_b - n_c/n_d = 1.733 - 0.4089 = 1.324, \tag{7.21}$$

$$echj \quad 1 + n_c/n_d - n_h/n_j = 1.409 - 0.0252 = 1.384, \tag{7.22}$$

$$ghcd \quad 1 + n_h/n_j - n_c/n_d = 1.025 - 0.4089 = 0.6162. \tag{7.23}$$

Table 7.4. Fractional cycles summation values for Case 2.

Damage	Load Path			
Metric d_a^+	*ecab*	*facd*	*echj*	*ghcd*
0	0.6493	1.351	1.317	0.6830
0.2	0.6764	1.324	1.384	0.6162
0.4	0.7196	1.280	1.448	0.5517
0.65	0.8035	1.197	1.477	0.5233
0.8	0.8746	1.125	1.402	0.5984
0.9	0.9325	1.068	1.263	0.7367
1.0	1.0	1.0	1.0	1.0

As for Case 1, corresponding results may be found for the damage metric values of 0, 0.2, 0.4, 0.65, 0.8, 0.9 and 1.0. The results are shown in Table 7.4.

Comparing the results in Tables 7.3 and 7.4 it is clear that the trends in the fractional cycles summations are similar for both types of partitioning.

7.4. Multiple Stress Range Changes

The above discussion is concerned with the effects of single stress range changes. In practice, interest extends to cases of multiple changes. Such cases may be treated by applying the foregoing results in Tables 7.3 and 7.4 and applying a linear fractional cycles summation rule. Other damage levels may be treated by interpolation of the tables.

Examples of two and three stress range changes are outlined below. Consider first the two-step change for the damage levels shown in Figure 7.5(a). (It will be recalled that $s_1 = 132$ MPa and $s_2 = 222$ MPa.)

Case 1. Data.
For the *carud* path in Figure 7.5(a), the fractional path length, t_1, is given by

$$t_1 = n_c/n_d + (n_r - n_a)/n_b + (n_d - n_u)/n_d. \tag{7.24}$$

Now applying Case 1 results above,

$$n_c/n_d = 0.1784; \quad n_u/n_d = 0.5630; \quad n_a/n_b = 0.5488; \quad n_r/n_b = 0.8187.$$

Substituting these values into Equation (7.24) yields $t_1 = 0.8854$.

Again in Figure 7.5(a), consider the path *acurb*. The fractional path length, t_2 in this case is given by

$$t_2 = n_a/n_b + (n_u - n_c)/n_d + (n_b - n_r)/n_b. \tag{7.25}$$

Substituting the given values into Equation (7.25) yields, $t_2 = 1.115$.

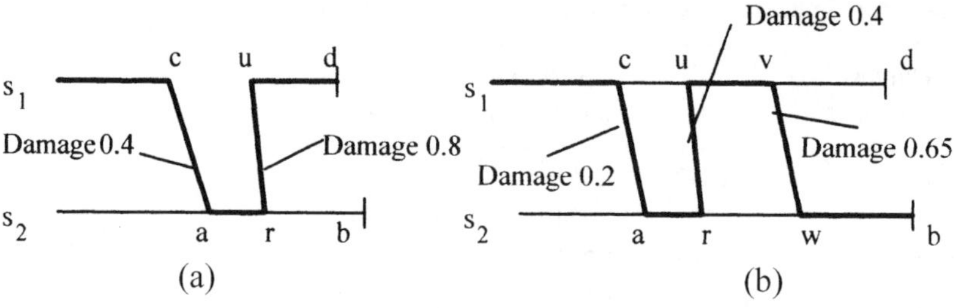

Figure 7.5. Two and three-step changes in stress range.

Consider next the three-step change in Figure 7.5(b). The fractional path length, t_1, for the path $caruvwb$ is given by

$$t_1 = n_c/n_d + (n_r - n_a)/n_b + (n_v - n_u)/n_d + (n_b - n_w)/n_b. \tag{7.26}$$

Again using the Case 1 results from Section 7.3.2

$$n_a/n_b = 0.4493; \quad n_r/n_b = 0.5488; \quad n_w/n_b = 0.7047,$$

$$n_c/n_d = 0.1004; \quad n_u/n_d = 0.1784; \quad n_v/n_d = 0.3659.$$

Substituting these values into Equation (7.26) yields, $t_1 = 0.6827$.

Evaluating the alternative path $acurwvd$, the path length, t_2, is given by

$$t_2 = n_a/n_b + (n_u - n_c)/n_d + (n_w - n_r)/n_b + (n_d - n_v)/n_d. \tag{7.27}$$

Substituting the given values into Equation (7.27) yields $t_2 = 1.317$.

Case 2. Data.

The same two and three-step paths are now considered again, but with Case 2 data.

Path $carud$ (Equation (7.24)).

In this case the appropriate values are:

$$n_a/n_b = 0.7918; \quad n_r/n_b = 0.9251,$$

$$n_c/n_d = 0.5114; \quad n_u/n_d = 0.7997.$$

Substituting these values into Equation (7.24) yields $t_1 = 0.8450$.

The alternative path is $acurd$ (Equation (7.25)). Substituting the values yields $t_2 = 1.155$.

Again, for the three-step path $carnvwb$ (Equation (7.26)

$$n_c/n_d = 0.4089; \quad n_u/n_d = 0.5114; \quad n_v/n_d = 0.6762,$$

$$n_a/n_h = 0.7325; \quad n_r/n_h = 0.7918; \quad n_w/n_h = 0.8727.$$

Substituting into Equation (7.26) yields $t_1 = 0.7603$.

For the alternative path, $acurvwd$ (Equation (7.27)), substitution gives $t_2 = 1.240$.

In the above examples the order of the path length values is the same, independent of which case data is used.

7.5. Remarks

In the FL treatment of cumulative fatigue damage described above no mathematical modelling is used and no material properties are prescribed. Only the broad observation is used that in high-endurance fatigue the stress range falls as the number of cycles to failure increases. There is also the postulate that the incipient microstructural damage has a similar pattern, but at lower stresses. The conclusions are in the form of fuzzy subsets and the implied tendencies (represented by the EFNs) are expressed as numerical values of the number of cycles.

On this basis the cumulative fatigue damage for decreasing and increasing stress ranges is considered. It is found that there is a tendency for the cumulative fractional cycles summation to failure to be less than unity for an increase in stress range, whereas the summation tends to be greater than unity for a decrease in stress range. This pattern appears to be persistent, even if the conditions are changed. This portion of the investigation is conducted in terms of dimensionless quantities, which means that the results are independent of scale. They are quite general and have significant design implications.

The methods do not depend upon having large numbers of experimental data available and are thus also suitable for application to model and prototype testing.

Chapter 8
Reliability Assessment

Reliability may be defined as the normalised frequency that a system, equipment or component can be expected to deliver its specified performance at a given point in "time". "Time" being any convenient independent metric of the system's service life, such as distance travelled, operational hours or cumulative product volume. Whether the performance satisfies the service demands depends upon whether or not the system is well matched to the requirements of the client and the environmental conditions. In practice, reliability cannot be entirely divorced from the competence of the operating and maintenance staff, although sometimes there is no information about this aspect at the design stage. For process plant the performance may be in terms of product quality and production rate, whilst for a railway system it may be in terms of the number of journeys per annum and the punctuality. Reliability concepts may also be applied to intangible services such as health care or postal services, for example. Since the reliability of a system is constantly changing in time, it is important that any expression of reliability should always be explicitly associated with its time.

The analysis of data to assess system reliability may be conducted at any of three levels:

(i)   system level, in which the boundary is a whole process or plant;
(ii)   equipment or assembly level, in which the boundary is a subsystem;
(iii)   component level, in which the boundary is a single component or non-repairable item.

Reliability has important implications for health and safety as well as for commercial aspects of an operation. There may be legal penalties in some circumstances and also a substantial value-at-risk. At the design stage a fault mode and effects analysis (FMEA) may be performed to establish the effects of potential fault modes at different levels within the system. Fault tree analysis (FTA) is a top-down approach to system failure analysis, starting with a system failure mode and tracing the causes down to lower levels ending with component level if required. It is possible and advisable to initiate a FTA at an early stage in the design to establish and update its reliability status as the design is developed.

A system may fail by deterioration of its critical parts over time or by sudden failure of a critical part. The former may be corrected to a certain extent by any available internal control. The latter by statistical process control and maintenance intervention (see Chapter 9).

8.1. Basic Formulae

8.1.1. HAZARD RATE AND RELIABILITY FUNCTIONS

In reliability data analysis, a hazard rate function, $z(t)$, is defined by

$$z(t) = -\mathrm{d}\ln r(t)/\mathrm{d}t, \tag{8.1}$$

where the reliability function $r(t)$ is related to the cumulative failure function $f(t)$ by

$$r(t) + f(t) = 1. \tag{8.2}$$

Note that in the literature the reliability function and the cumulative failure function are invariably denoted by capital letters, but here capital letters are again reserved for sets and lower case letters denote UDs.

The initial and final conditions are: $f(t_o) = 0$, where t_o is the time at the initiation of operations and $f(\infty) = 1$.

Let $\phi(t' - t_o)$ be the failure frequency distribution in the interval (t_o, t) where t' is non-current time. Then

$$f(t) = \int_{t_o}^{t} \phi(t' - t_o)\, \mathrm{d}t. \tag{8.3}$$

Integrating Equation (8.1)

$$\int_{r_o}^{r} \mathrm{d}\ln r(t) = -\int_{t_o}^{t} z(t')\, \mathrm{d}t'. \tag{8.4}$$

Hence,

$$r/r_o = \exp\{h(t_o) - h(t)\}, \tag{8.5}$$

where

$$h(t_o) - h(t) = -\int_{t_o}^{t} z(t')\, \mathrm{d}t'.$$

If $r_o = 1$, then

$$r = \exp\{h(t_o) - h(t)\}. \tag{8.6}$$

For a constant hazard rate function, $z(t) = 1/\lambda$ the reliability function may be expressed as

$$r(t) = \exp\{(t - t_o)/\lambda\}. \tag{8.7}$$

A similar expression to (8.7) is frequently used in the engineering literature.

8.1.2. FAILURE PROBABILITY DISTRIBUTIONS

The conventional approach to the assessment of reliability data is through the fitting of a continuous or discrete failure probability distribution model with sufficient disposable constants to laboratory or field data. BS 5760: Part 2 describes this method. Continuous distributions are applicable to the majority of engineering data, discrete distributions are appropriate for a given number of "success" or "failure" tests. The most familiar of the continuous distributions are the normal, exponential, log normal and gamma forms. The normal (Gaussian) is frequently applied and has two disposable constants; it provides an "S" shaped cumulative failure function and a rising hazard rate function with time. It is however rather unwieldy to manipulate. The exponential function with one disposable constant is easy to manipulate and is therefore popular; the cumulative failure function rises monotonically with time, but at a diminishing rate, whilst the hazard rate function is constant. Its proper application is to cases of constant failure rate. The log normal and gamma distributions are both more complicated forms and are less frequently used in practice; both have two disposable constants.

8.1.3. THE WEIBULL FORMULA

The most widely published method of reliability assessment in the engineering literature is based upon the Weibull empirical formula

$$f(t) = 1 - \exp\{-[(t - \alpha)/\eta]^{\beta}\}, \tag{8.8}$$

where $\alpha < t$.

Equation (8.8) represents a three-parameter model in which η is a scale parameter or characteristic time (the time required for approximately 63.2% of the original component population to have failed), β is a shape parameter and α is a delay period before any failures begin. There is computer software available for the evaluation of the constants in Equation (8.8). More traditionally, the data may be plotted on commercially available Weibull charts and this has some advantages in providing a visual appreciation of the adequacy of the formula and also of the occurrence of outliers in the data. The Weibull chart has double logarithmic versus logarithmic axes; such scaling will often obscure physical features and is efficient at producing an apparently linear graph. Competing failure modes are one possible cause of non-linearity on a Weibull chart. The two disposable constants β,

η are obtainable directly from the chart. For an apparently linear Weibull plot, α is assumed to be zero.

For small values of $(t - \alpha)/\eta$ a plot of $\{f(t)\}^{\beta}$ will provide an estimate of α as $f(t) \to 0$.

As time progresses the artefacts would usually pass through different failure modes, but these are not revealed and differentiated in the Weibull chart. The evaluated constants therefore do not have any fundamental significance.

8.2. Piecewise Constant Hazard Rate Function (PCHRF)

The hazard rate function defined by Equation (8.1) which is constant implies a loglinear reliability function, Equation (8.7). If the artefact has several dominant failure modes then these would show as several near loglinear reliability functions within the dominant zones.

Consider Equation (8.7), then in the more general case of n dominant modes, in the ith mode at time t the locally appropriate reliability function is

$$r_i(t) = \exp\{(t - t_{io})/\lambda_i\}, \quad i = 1, 2, \ldots, n. \tag{8.9}$$

Let the $r_i(t)$ define the support (span) of n interpenetrating fuzzy sets, R_i, on the t UD. At any time t, the membership functions of the adjacent R_i are μ_i and μ_{i+1}. The resulting reliability function is defined as

$$r(t) = \mu_i^m r_i + \mu_{i+1}^m r_{i+1}, \tag{8.10}$$

where $0 \leq \mu_i \leq 1$ and usually

$$\mu_i + \mu_{i+1} = 1. \tag{8.11}$$

Equation (8.10) represents the transition functions between the dominant modes of failure. Thus a smooth continuous representation of the reliability function is found by patching together the functions of the dominant modes. As many R_i may be taken as are required for the desired accuracy of portraying the physical data, and the number of parameters is not limited to three as is the case for the Weibull analysis. The index m in Equation (8.10) allows "tuning" of expression for the transition curve to pass through a fair point in the data.

8.3. Human Reliability Assessment

Human participation in system functioning is very common in control, maintenance and supervision not only in artificial, but also in health and environmental systems. A total assessment of system reliability cannot be complete without consideration of this factor. To include human reliability within the overall system reliability assessment requires its quantification. This is not so simply achieved for several reasons:

(i)   there are significant subjective aspects to all data;
(ii) there is a scarcity of published field data;
(iii) poor performance may be suppressed for publicity reasons;
(iv) simulator studies may be unrepresentative.

Automatic control and expert systems can provide limited alternatives to some of the roles of the human, but there are reasons such as costs why they are not found more widely. There are also cases where the versatility and intuitive judgement of the human is preferred, hence the popularity of the manual car gear box amongst some drivers.

8.3.1. TYPES OF ERRORS

Human reliability depends upon the frequency of errors in the number of opportunities. BS 5760: Part 2 identifies four main types of human error:

(i)   Omission: failure to execute the necessary action;
(ii) Commission: inadequate action, too much or too little force, inaccurate action, ill-timed action, wrong action sequence;
(iii) Extraneous actions: unneeded action;
(iv) Corrections: lost error-correction opportunities.

The above are not mutually exclusive categories. Item (i) could be deemed to include item (iv) for example. But the salient errors found in practice could be readily fitted into one of these broad categories.

8.3.2. ERROR PROBABILITY

The human error probability (HEP) must be quantified for integration into a total system reliability assessment. This is defined as the normalised frequency of a particular type of error in a given period of time,

$$\text{HEP} = n_e/n_o,$$

where n_e is the number of errors in a given period and n_o is the maximum number of opportunities for the same error in the same period. In reality it is difficult to obtain proper field data of this type as discussed above, but BS 5760: Part 2 provides a table of generic HEPs as shown in Table 8.1. This provides a general basic guide to failure probabilities, dividing the range of task difficulties into five categories ranging from easy familiar tasks at one end to demanding unfamiliar tasks.

The right-hand column of the table classifies the probabilities as fuzzy sets.

HEP values are obtained almost entirely from simulator and laboratory experimental tests, there is very little field data available, because it is very difficult to gather. Organisations would be reluctant to release such information even if it were

Table 8.1. Task categories and generic HEPs.

Task Category Fuzzy Set	Failure Probability pu	
1. Simple, frequently performed tasks, minor mental demands.	0.001	
2. Moderate difficulty, less time, more mental demand.	0.01	
3. Complex task, strong mental demand.	0.1	Medium
4. Higher complexity, very strong mental demand.	0.3	High
5. Limiting mental demand, unfamiliar task.	1.0	

available. The available data are therefore not entirely realistic and representative of working conditions. The sharp numerical boundaries of the categories in Table 8.1 express a precision which is not realistic and the position of the boundaries is also arbitrary. A more natural partitioning is that given by the fuzzy sets in the final column of the table. This properly describes the uncertainty in the membership of the interpenetrating sets rather than present an apparently well defined set of categories.

In a specific case the possible error is identified from an action flow diagram for a particular task. The skill level of the operator and also the equipment available will affect the HEP in a given case.

8.3.3. THE EFFECT OF ENVIRONMENT

Where a similar task may be carried out in a variety of environments ranging from benign to adverse, this would have an effect on the HEP, given that all other factors remain the same. It is possible to create a framework which reflects the environmental effects.

For the present purpose it will be adequate to categorise the environment conditions as: Poor (PR), Fair (FA), Good (GD), Excellent (EX). This partitions the UD into four interpenetrating fuzzy sets as shown in Figure 8.1(a). The UD is scaled 0–1.0. The task grades shown in Table 8.1 can be retained and represented by five fuzzy sets labelled G1 to G7 on the UD which is also scaled 0–1.0, shown in Figure 8.1(b). The failure probability in Table 8.1 spans a thousand fold and therefore the UD is placed on a logarithmic scale and is partitioned into five sets similar to the spacing in Table 8.1. These sets are given linguistic terms: Very Low (VL), Low (LO), Medium (ME), High (HI) and Very High (VH). The sets are illustrated in Figure 8.2.

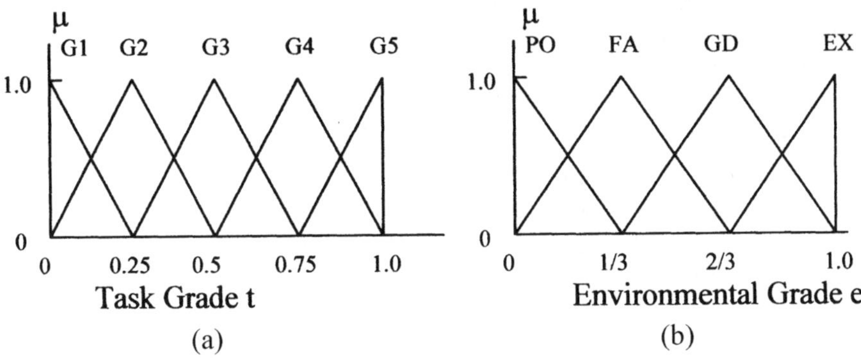

Figure 8.1. Partitioning of the UDs. (a) Task grade. (b) Environmental grade.

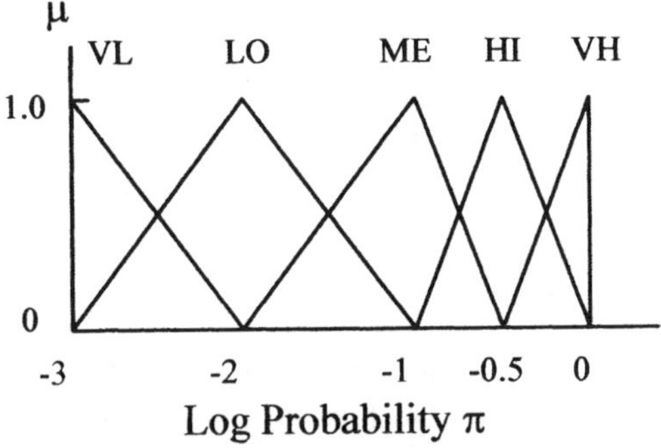

Figure 8.2. Partitioning of the log probability UD.

The membership functions corresponding with the partitioning illustrated in Figures 8.1(a) and (b) and 8.2 are given below.

Task grade	$0 \le t \le 0.25$	$0.25 \le t \le 0.5$	$0.5 \le t \le 0.75$	$0.75 \le t \le 1.0$
μ_{G1}	$1 - t/0.25$			
μ_{G2}	$t/0.25$	$2 - t/0.25$		
μ_{G3}		$(t - 0.25)/0.25$	$3 - t/0.25$	
μ_{G4}			$(t - 0.5)/0.25$	$4 - t/0.25$
μ_{G5}				$(t - 0.75)/0.25$

Environment	$0 \le e \le 1/3$	$1/3 \le e \le 2/3$	$2/3 \le e \le 1.0$
μ_{PO}	$1 - 3t$		
μ_{FA}	$3t$	$2 - 3t$	
μ_{GD}		$(t - 1/3)/1/3$	$3(1 - t)$
μ_{EX}			$(t - 2/3)/1/3$

Table 8.2. Relationship array for HER failure probability.

	G1	G2	G3	G4	G5
PO	ME	ME	HI	VH	
FA	LO	ME	ME	HI	VH
GD	VL	LO	ME	HI	VH
EX	VL	VL	ME	HI	VH

Probability	$-3 \leq \pi \leq -2$	$-2 \leq \pi \leq -1$	$-1 \leq \pi \leq -0.5$	$-0.5 \leq \pi \leq 0$
μ_{VL}	$-2 - \pi$			
μ_{LO}	$3 + \pi$	$-1 - \pi$		
μ_{ME}		$2 + \pi$	$-1 - 2\pi$	
μ_{HI}			$(1 + \pi)/0.5$	-2π
μ_{VM}				$(0.5 + \pi)/0.5$

where $\pi = \log p$.

The corresponding relationship array is created by an induction process based upon searching the organisation's free knowledge base to find consensus judgement. A relationship array of the type shown in Table 8.2 is formed.

The above is a broad treatment of environmental influence and could be applied, for example, to 24 hour outdoor maintenance.

EXAMPLE 8.1. A certain task is assigned a grade of 0.4 and the environmental grade is judged to be 0.6. Find the probability of failure. Compare this probability with that of the same task carried out in an environment grade of 0.3.

Solution. In the first case the membership values are:

$$\mu_{G2} = 2 - 0.4/0.25 = 0.4; \quad \mu_{G3} = 0.6;$$

$$\mu_{FA} = 2 - 3 * 0.6 = 0.2; \quad \mu_{G2} = 0.8.$$

The FL proposition is

IF T	AND E	THEN Π	MIN	CONSEQUENCE
G2	FA	LO	0.4, 0.2	0.2LO
G2	GD	LO	0.4, 0.8	0.4LO
G3	FA	ME	0.6, 0.2	0.2ME
G3	GD	ME	0.6, 0.8	0.6ME

The conclusion (Π) is the union of the consequences,

$$\Pi = 0.4\text{LO} \cup 0.6\text{ME}.$$

This shows that the dominant probability element lies in the medium set but with a lesser element in the low set.

To find a deterministic equivalent, note that $\mu_{ME} = 2 + \pi$. Hence, $0.6 = 2 + \pi$. Therefore, $\pi = -1.4$ or $\log p = -1.4$; $p = 0.04$.

In the second case where the environmental grade falls to 0.3, the task membership values remain as before. The new environmental membership values are $\mu_{FA} = 3 * 0.3 = 0.9$; $\mu_{PO} = 0.1$.

In this case the FL proposition is

IF T	AND E	THEN Π	MIN	CONSEQUENCE
G2	FA	ME	0.4, 0.9	0.4ME
G2	PO	ME	0.4, 0.1	0.1ME
G3	FA	ME	0.6, 0.9	0.6ME
G3	PO	HI	0.6, 0.1	0.1HI

The conclusion is the union of the consequences

$$\Pi = 0.6\text{ME} \cup 0.1\text{HI}.$$

The dominant probability lies in the medium set, as before, with a minor element in the high set.

A deterministic equivalent may be found using the membership weighted values

$$\text{def } \Pi = [0.6(-1) + 0.1(-0.5)]/0.7 = 0.928.$$

Therefore, $\log p = -0.928$ or $p = 0.118$.

According to this treatment, the probability of failure increases from 0.04 to 0.118 when the environmental grade falls from 0.6 to 0.3. These probability values are indicative values of tendencies, the full solutions are given by the fuzzy sets, Π.

8.4. Data Processing

There is no deterministic theory of failure or reliability for systems, equipment or components. Reliability assessment depends upon laboratory tests or in-service field data and curve fitting using statistical or empirical models with sufficient disposable constants. Assessments rely both on the quality of the data and its subsequent preparation. One of the reliability formulae is then applied to model reliability or failure correlations and as a basis for performance predictions. For this purpose the failure data must be a proper representative sample of a population operating under similar conditions to the population as a whole. The model can then be used to make deductions about operational matters such as maintenance strategy, based upon the same failure criterion, which needs to be well defined.

Cumulative failure metric

Let t_i be the failure time of i samples of a total of n items and let t_n be the failure time of the nth item. It may seem reasonable to quantify the cumulative failure metric (f) as $f(t_i) = i/n$. For "small" sample sizes $(n < 100)$, this formula gives an overestimate of $f(t_i)$, which becomes more serious as the sample size decreases. This error is largely overcome by the application of the widely used Bernard's approximation to median rank,

$$f(t) = (i - 0.3)/(n + 0.4). \tag{8.12}$$

This is a suitable cumulative failure metric for most engineering data.

Censored data

There are two main types of failure and reliability data: one in which all the sample items have failed, which is called complete, it would typically result from laboratory tests. The other is the 'censored' data, which may be singly or multiply censored data, and it might result from either laboratory or field service observations. Singly censored data means data where all survival ages are equal to or larger than the largest time to failure in the data. This means that if 15 of a sample of 20 items have failed when the test is terminated, then the sample size is 20 and not 15.

Multiply censored data means that some of the items have been withdrawn from the test programme for reasons other than the particular type of failure under observation. This happens more frequently in field than in laboratory data. The cumulative failure value is calculated from estimates of the cumulative hazard function $\{h^{\#}(t_i)\}$. At the point at which failure occurs the sample estimate of the hazard rate function, $z^{\#}(t_i)$, is

$$z^{\#}(t_i) = \text{Number of failures occurring at time } t_i \text{ / Number of survivors at time } t_i.$$

The value of $h^{\#}(t_i)$ is the given by

$$h^{\#}(t_i) = \sum_i z^{\#}(t_i). \tag{8.13}$$

Finally, the estimate of the cumulative failure function is

$$f^{\#}(t_i) = 1 - \exp -h^{\#}(t_i). \tag{8.14}$$

EXAMPLE 8.2. Failure data for a sample of 10 items is given in BS 5760: Part 2. Reliability values calculated from the data are given in the following table. The original failure data were based upon Bernard's formula, Equation (8.12).

Failure number	Ranked time to failure hrs.	Reliability
1	300	0.933
2	410	0.838
3	500	0.741
4	600	0.645
5	660	0.548
6	750	0.452
7	825	0.355
8	900	0.259
9	1050	0.162
10	1200	0.067

Determine whether the data can be represented by a PCHRF and if so, evaluate the constants. Also,

(i) find the expected % failure at 400 hours life,
(ii) the time for 30% failures and for 63.2% failures,
(iii) the hazard rates.

Solution. It is noted that the data are complete with no censoring. The data are illustrated (the accuracy is limited by the software) on a loglinear graph in Figure Ex 8.1. It is apparent on this graph that the data may be reasonably fitted by three straight lines. An analysis of these graphs gives the following relations:

$$\text{Curve 1} \quad r_1 = \exp\{-(t - 229)/1025\},$$

$$\text{Curve 2} \quad r_2 = \exp\{-(t - 325)/567],$$

$$\text{Curve 3} \quad r_3 = \exp\{-(t - 650)/198\}.$$

The intersection of curves 1 and 2 is at $t = 443.8$ hrs and the intersection of curves 2 and 3 is at $t = 824.2$ hrs. These intersections will be taken as the cross-over points of the fuzzy reliability sets. The transition curves between curves 1 and 2, and between curves 2 and 3 will be defined by the partitioning of the reliability fuzzy sets on the t UD. It is noted from Figure Ex 8.1 that the mutual interpenetration between the fuzzy reliability sets is comparatively small and adequately covered by a value of ± 40 hrs about the cross-over points. The partitioning of the UD is shown in Figure Ex 8.2.

The membership functions corresponding with Figure Ex. 8.2 are given below.

	$0 \le t \le 404$	$404 \le t \le 484$	$484 \le t \le 784$	$784 \le t \le 864$	$864 \le t$
μ_1	1	$6.05 - 0.0125t$	0	0	0
μ_3	0	$(t - 404)/80$	1	$10.8 - 0.0125t$	0
μ_3	0	0	0	$(t - 784)/80$	1

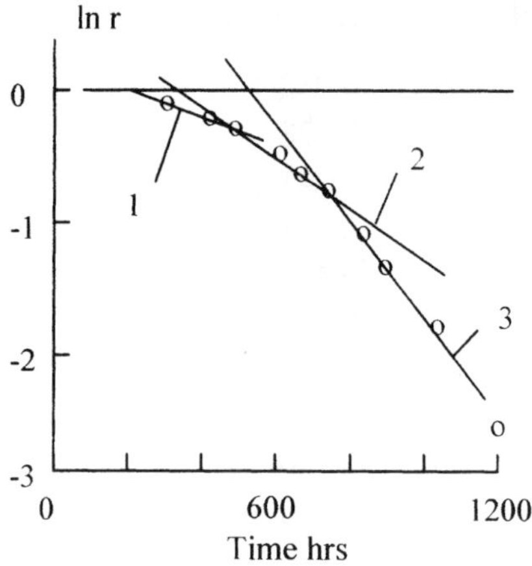

Figure Ex 8.1. Log r as a function of time.

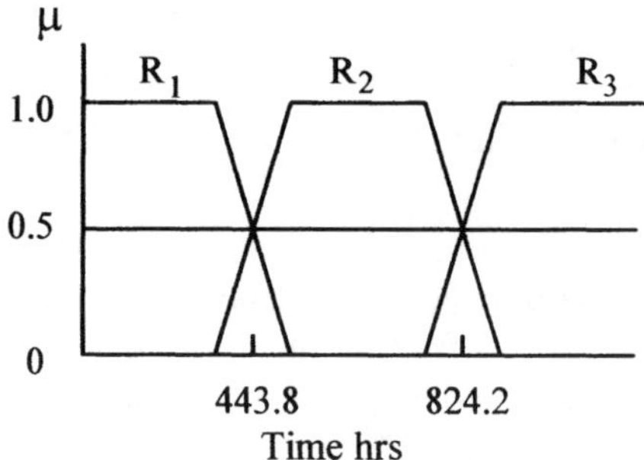

Figure Ex 8.2. Partitioning of the t UD.

Values of t have been rounded off to the nearest whole number whilst retaining sufficient accuracy.

At the intersection of curves 1 and 2, $\mu = 0.5$ and $r_1 = r_2$, hence from Equation (8.10)

$$r(t) = 2(0.5)^m r_1.$$

Solving this equation for m would provide a transition curve which passes through the intersection point, which lies above the required curve. Therefore a fair point on the curve is selected as $0.98r(t)$. With this substitution the solution of the equation gives $m = 1.029$. Hence the first transition curve is

$$r(t) = \mu_1^{1.029} r_1 + \mu_2^{1.029} r_2.$$

Similarly, the second transition curve is

$$r(t) = \mu_2^{1.029} r_2 + \mu_3^{1.029} r_3.$$

(i) Noting the partitioning of the t UD, the proportion of failures after a 400 hr life may be found from

$$r_1 = \exp\{-(t - 229)/1025\},$$

i.e.

$$r_1(400) = \exp\{-(400 - 229)/1025\}, \quad r_1 = 0.8462 \quad \text{and} \quad f_1 = 15.38\%.$$

(ii) Figure Ex 8.1 indicates that the time for 30% failures may be found from

$$r_2 = \exp\{-(t - 325)/567].$$

Thus

$$t = 325 - 567 \ln r$$

$$= 325 - 567 \ln 0.7$$

$$= 527 \text{ hrs.}$$

Similarly for 63.2% failures

$$r_3 = \exp\{-(t - 650)/198\}$$

or,

$$t = 650 - 198 \ln r,$$

i.e.

$$= 650 - 198 \ln 0.368$$

$$= 849 \text{ hrs.}$$

The BS 5670:Part 2 analysis is on the basis of the Weibull formula Equation (8.8) which provides the following values: (i) 15% (about), (ii) 550 hrs and 820 hrs. These are quite similar to the values found by the above PCHRF model. There is a difference in that the Weibull analysis gives a zero delay time ($\alpha = 0$), whereas the above method gives an initial time delay of 229 hrs. The double log versus log scale of the Weibull chart is effective in masking detail.

(iii) The hazard rates for curves 1, 2 and 3 are $9.755 * 10^{-4}$ hrs^{-1}, $1.764 * 10^{-3}$ hrs^{-1} and $5.043 * 10^{-3}$ hrs^{-1}. These are found as the respective values of the $1/\lambda_i$.

8.5. Remarks

Reliability assessment is entirely dependent upon fitting data correlations from laboratory or field observations using one of the various standard forms of probability distribution. Most of these are mathematically cumbersome and have either two (usually) or three disposable constants for curve fitting. The empirical Weibull formula has three, but in many practical cases one of these appears to be zero. The double log versus log Weibull plot is very effective in producing linear plots, but obscures physical features.

A different approach is to define locally constant hazard rate functions, which imply reliability functions each with two constants to be determined from the physical data. This is a flexible treatment in that as many hazard rate functions as are required may be used. In practice, two or three are sufficient to accurately define the physical data. The local reliability functions are pasted together adopting FL precepts in the transition zones enabling an extensive relationship to be formed covering the whole of the physical data range.

Human reliability assessment depends for the most part upon laboratory data, which is only partially representative of actual field conditions. There are cases too where laboratory simulations are not available and data must be synthesised using generic data. The conventional way of representing the information in mathematical and numerical form implies a degree of precision which is not supported by the underlying uncertainty in the information. A more natural form is in linguistic terms, which does not over represent the inherent lack of precision. This is exemplified by the FL treatment of the effect of environmental conditions.

Chapter 9
Process Control

The two major types of problem in control are those of tracking and regulation. They are not different fundamentally, only in emphasis. Process control is of the regulation type and aims at achieving and maintaining a product stream within specification. Whether it be a stream comprising components, assemblies, bagged, packed or containerised solids, fluids or particulate matter, this is not easy to achieve. The product stream is subject to a range of influences such as:

  (i)   input materials quality,
 (ii)   process function and state variability,
(iii)   environmental conditions,
 (iv)   operator reliablity,
  (v)   reliability and quality of support services such as maintenance.

Statistical process control (SPC) is a well-known and widely publicised tool which is employed in pursuance of product specification compliance through the regulation of the production process. It will be obvious from the above list that SPC is generally effected by external intervention, unlike automatic process control (APC), which is generally a closed-loop and on-line subsystem of the process. These characteristics are incorporated into the control strategy described below.

Control charts play a prominent role in SPC applications and although they originated in the light engineering sector, their use has since spread to other areas of production and processing, such as the chemical industry and to machinery condition monitoring. The control chart is conventionally demarcated by upper and lower control limits (UCLs and LCLs respectively), which are often symmetrically placed about a central value line. A refinement of this demarcation is by subdividing each of the two intervals between the central value line, the UCL and the LCL, into three equally spaced zones.

In the application of control charts it is emphasised that each case needs to be considered in detail and treated on its own merits to determine the appropriate form of control chart. This is also true in the wider context of SPC-APC applications. At the commissioning stage it must be established in the first place, whether or not the system is capable of providing an output of the requisite specification at the desired rate. If it is, then some of the control function may be exercised by APC. Those parts of the control function that cannot be managed under APC, because of the type or range of action required, must be managed under SPC and external intervention such as corrective maintenance or change of inputs for example. APC may, or may not, share a common sampling method with the associated SPC.

In batch processing, steady state is not achieved and therefore successive samples are not drawn from the same unit and are not statistically uniform, though sample analysis may, of course, be compared with a specification. However, statistical conclusions drawn under such circumstances are of questionable validity.

For any control system of whatever type, the time scale of the control (including the sampling system) response must be of smaller order of magnitude than that of the production process for it to be effective. The process may need to be stopped for SPC intervention, but APC is normally on-line. Although it is known that conventional linear APC can sometimes be applied to non-linear systems by tuning using the Ziegler-Nichols rules (see standard control texts), a fuzzy logic representation is used in this work, providing a fuzzy automatic process control (FAPC). Such a representation may be applied even when no mathematical model of the system is known.

9.1. Basic SPC Concepts

If a system is under control it means that its output is statistically uniform and that the product is all from the same universe. This does not mean that the product mostly complies with its quality specification. The quality will be in terms of a single critical parameter target value (with a tolerance range), typically a dimension, strength, weight or electrical resistance, for example. The observed average of the output samples may not coincide with the target values or, even if it does, the scatter in the values may be too great. The level of process control must match the quality specification. If the former type of fault is noted, it may be corrected by process adjustment under APC or SPC, but if the latter type of fault occurs it may entail more fundamental action with off-line intervention, perhaps up to system redesign. An alternative is to amend the level of control to match the specification.

In any application of SPC the hypothesis being tested is that any detected variation in the data is only due to stochastic processes. Two types of variation maybe observed in sample data:

(i)   Chance causes. These are due to unassigned events within the process or small variations in environmental conditions to which the process is sensitive, variations in inputs or in operator actions.

(ii) Assignable variations. These are due to accountable causes and usually produce identifiable patterns on the control chart. They are produced, for example, by a process deterioration, mechanical faults, a change in source of raw material or process operator fatigue.

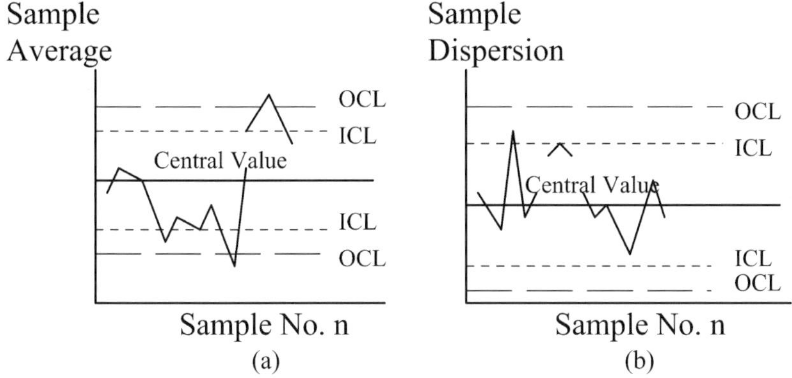

Figure 9.1. Typical control charts. (a) Sample averages. (b) Sample dispersions.

The effects of both types of variations are additive. In the unassignable chance causes category, part of this may be due to random effects which can be termed "noise". It is well known that integration reduces the effect of signal noise and, similarly, the effect of averaging sample values of, say, 5–10 production units will assist in mitigating the effects of random or periodic variations. Clearly, the larger the sample size, the more effective is the reduction, but subject to control system and economic efficiency. SPC cannot be applied to chance causes. In the case of non-random variations in the system output, the identification and classification of causes are important. Frequently, in system design a fault tree analysis (FTA) is conducted which provides vital information on the most likely causes of output variation. If this is not available, the first step is to create those FTAs that are relevant to the process. The FTAs will provide the most probable cause and effect relationships which will guide process control.

9.2. Control Charts

Normally, two types of chart are constructed: (a) a chart for sample averages and (b) a chart for dispersion, which may be either standard deviation or range. Examples of these two control chart types are illustrated in Figures 9.1(a) and (b). Figure 9.1(a) is for sample average and shows the outer control limits (OCLs) and inner control limits (ICLs). For a Gaussian (normal) probability distribution the OCLs are at $x_o = x_a \pm 3.06\sigma/n^{1/2}$, while the ICLs are at $x_i = x_a \pm 1.96\sigma/n^{1/2}$, where σ is the standard deviation of the universe, x_a is the central value and n is the sample size. With these limits, 5% of the samples will lie outside the ICLs, but only 0.2% will lie outside the OCLs. Sometimes a simpler type of chart is used with only two control limits, called the upper and lower control limits (UCL and LCL respectively). These limits correspond closely with the position of the two OCLs.

Generally, the control limits may be expressed as:

Average:

$$\text{OCL:} \quad x_o = x_a \pm A(n)\sigma, \tag{9.1}$$

$$\text{ICL:} \quad x_i = x_a \pm B(n)\sigma. \tag{9.2}$$

Range:

$$\text{OCL:} \quad r_o = r_a \pm C(n)\sigma, \tag{9.3}$$

$$\text{ICL:} \quad r_i = r_a \pm D(n)\sigma. \tag{9.4}$$

There are tabulated values of $A(n)$, $B(n)$, $C(n)$ and $D(n)$ or the Gaussian distribution and also for $A(n)$ and $C(n)$ for different types of probability distribution.

9.3. Interpretation of Observations

The simplest rule of SPC is that s sample values lying outside the control limits indicates a process out-of-control condition. If the causes of the observed fluctuations about the central values are stochastic, then sample values have a 50% probability of being on either side of the average. If a number of consecutive values fall on one side of the average, then there is a probability that there is some deterministic cause. For example, if seven consecutive values lie on one side of the average, the probability of this is 0.78%, which shows that this is an unlikely event. The probabilities of other patterns, based upon a Gaussian distribution can also be found. These are further indicators of process out-of-control conditions.

If the sample average continues to be biased on one side of the control chart whilst the dispersion remains unchanged, then corrective action is indicated. The process itself is however, likely to be controllable. Figure 9.1 shows control charts for a process that is out of control, but controllable. If, however, the dispersion is biased on the positive side, but the sample means remain symmetrical, then a more fundamental process fault has occurred which is unlikely to be controllable by APC, but requires off-line correction by repair or replacement maintenance, or perhaps process modification. If the dispersion is biased on the negative side of the average, it may be caused by a reduction in accuracy of the sampling and inspection procedure, which again may require off-line correction. It is clear from the above discussion that clusters of sample observations that are biased on one side or the other of the central value of the sample averages or dispersion indicate a process fault. This has been formalised by introducing zones within the OCLs at $\pm\sigma$, $\pm2\sigma$ and $\pm3\sigma$ on both the sample average and dispersion charts. For a Gaussian distribution, approximately 66.7% of the sample values lie in the 0–σ zone, 95% lie in the 0–σ zone and 99.75% lie in the 0–σ zone.

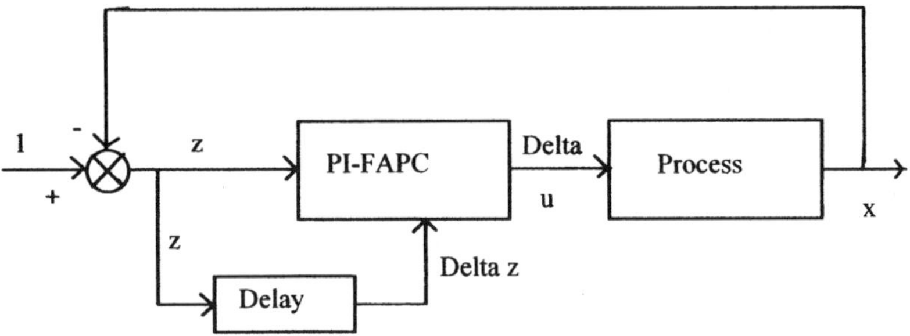

Figure 9.2. PI process control block diagram.

9.4. Fuzzy Automatic Process Control (FAPC)

For process regulation, proportional-integral (PI) control is commonly applied and this is the type that could be employed in the present context. It is emulated in fuzzy logic terms in the feed-back control described below. The block diagram in Figure 9.2 shows the principle of PI-FAPC. The basic theory is described elsewhere in the companion volume.

Tracking can also be achieved by the incorporation of a certain degree of derivative control, but with the penalty of increasing instability. There is no general theory of non-linear process control; however, FAPC as outlined below can be applied to any type of process, whether or not it is linear, or indeed has any known mathematical model. Sensitivity and stability are important concepts in APC and are discussed in standard control theory texts, however stability tends not to be a problem in PI control. There is no available stability theory for SPC or FAPC, each case must be studied on its own merits. If the sampling time and the control response time are together equivalent to n_c units of production, while the process response time is equivalent to n_p units of production, then the overall system response time is $O[(n_c+n_p)/q]$, where q is the system production rate. The sampling time depends upon the number of consecutive production units per sample and also the number of samples in a given control rule. An alternative to periodic sampling is to use rolling values of sample averages and dispersions.

Although both sample averages and dispersion are of interest in process control, it is noted above that out-of-control dispersion is most likely to be caused by a process fault requiring repair or replacement maintenance, which is out of the province of FAPC. Therefore, in the following discussion only control of the average value is considered.

At some point in the production, let the error in the specification of a critical quantity be z and its incremental change be δz. The resulting control signal is δu.

Representations of the partitioning of the z, δz and δu spaces are shown in Figures 9.3, 9.4 and 9.5 respectively. The z and δz comprise the control inputs. The QL categories are possible quality ranges, interpreting product faults as quality levels enables supplier penalties to be imposed.

Table 9.1. Typical PI controller relational base.

	δz					
z	**A**	**B**	**C**	**D**	**E**	**F**
QL1	L	K	K	K	J	J
QL2	K	K	K	J	J	J
QL3	K	J	J	J	J	H
QL4	K	J	J	J	H	G

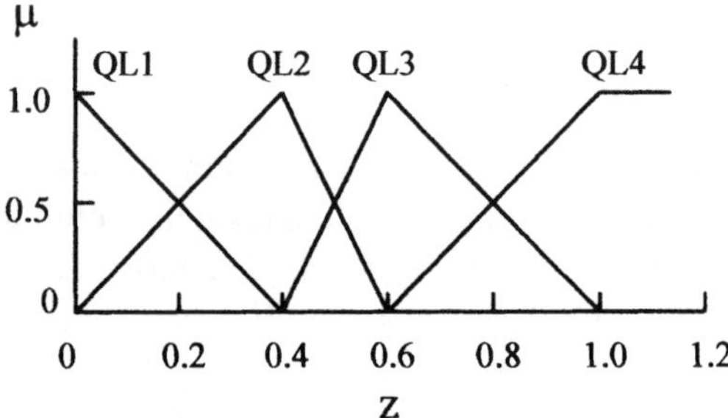

Figure 9.3. Partitioning of the z space.

A typical relational case for a PI-type controller output is shown in Table 9.1. The control process is governed by the operation of a fuzzy logic intersection form of proposition

IF Z   AND δZ   THEN U

The operation is similar to the intersection operation in two-valued logic. Since both Z and δZ yield two fuzzy sets each, there are four consequences, which, by logical union, comprise the fuzzy conclusion. The defuzzified numerical value of the conclusion provides the control input signal to the production process.

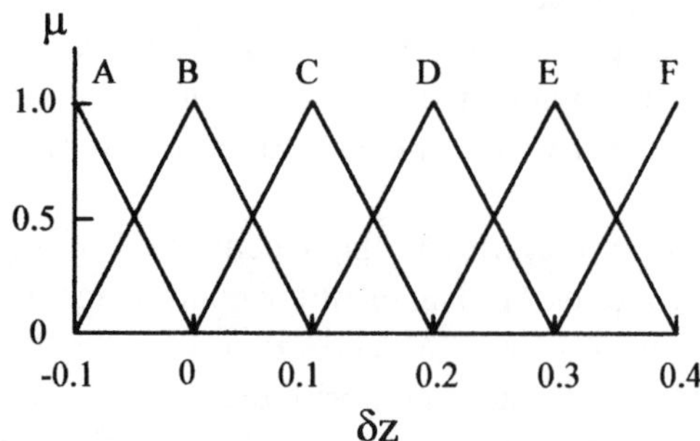

Figure 9.4. Partitioning of the δz space.

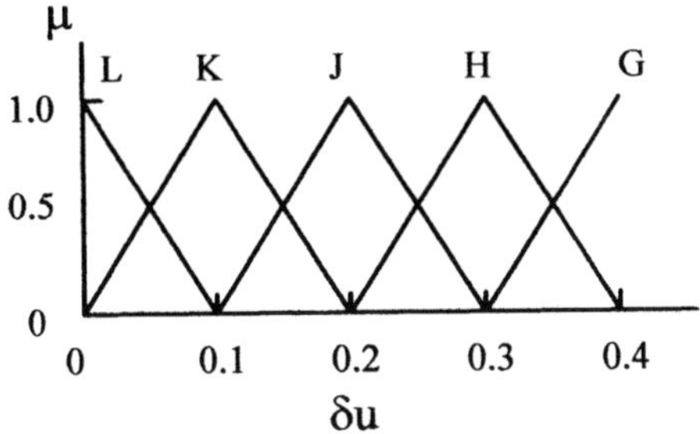

Figure 9.5. Partitioning of the δu space.

9.5. Control Strategy

Consideration is now given to the correlation of SPC and FAPC for out-of-control production process states to provide an overall process control strategy. Each critical output parameter must be treated individually, as described below.

There are four recognised types of control chart pattern, which are assumed to signal an out-of-control process state. These are:

(i)   Outliers: a single sample beyond the OCLs (or UCL and LCL);

(ii)   Trends: consecutively increasing or decreasing sample values;

(iii)   Biases: consecutive sample values, the majority of which are biased on one side or other of the central value;

(iv)   Oscillations: a repetitive pattern of consecutive sample values.

An outlier is an isolated value and an unsuitable candidate for FAPC, which essentially takes pre-emptive action on subsequent process output, on the basis of historical evidence. SPC with production halted is then the only possible course of action. Trends are caused by progressive process deterioration, environmental changes or operator fatigue, for example, and these may be compensated for to a limited extent by FAPC. Biases that are not preceded by a trend are typically caused by a change of operator, or in input specification, or a sudden shift in the process state or sampling system. In specific cases these may also be compensated to a limited extent by FAPC action, but more generally require external intervention. Whether oscillation can be compensated by FAPC action depends upon their periodicity and amplitude. They will be partially masked if rolling means samples are used.

The proposed strategy is summarised in Table 9.2. Since FAPC is preferred to SPC intervention it is possible to consider installation of FAPC for trends and biases, provided that it can be economically justified. If trends or biases become sufficiently large they become outliers, the process is stopped and control action transfers from FAPC to SPC. A correlated SPC-FAPC system is shown in the block diagram in Figure 9.6. The control system comprises four modules for sample

Table 9.2. SPC-FAPC control strategy for process patterns (i) to (iv).

Pattern	FAPC	SPC
(i) Outliers	No	Yes
(ii) Trends	Limited	Yes
(iii) Biases	Limited	Yes
(iv) Oscillations	No	Yes

analysis, signal discrimination for sorting data into FAPC or SPC processing, fuzzy control, an SPC expert function and an FTA knowledge base. The functions of these units are illustrated in Figure 9.6. The functions of the expert module for dispersion data are shown in some detail in Figure 9.7. This module operates with IF-THEN logic operations, e.g. for dispersion data:

IF BELOW LIMIT THEN CHECK SENSOR FUNCTIONS SENSITIVITY
IF ABOVE LIMITS THEN CHECK PROCESS ACCURACY COMPARED
WITH LIMITS

For central values the logic operations for a mechanical system might be:

IF BELOW LIMIT THEN CHECK SENSOR FUNCTIONS
IF ABOVE LIMIT THEN CHECK:
(1) PROCESS FRICTION
(2) LOOSE CONNECTION
(3) EXCESSIVE WEAR

The prioritisation would be obtained from the process fault tree analysis (FTA) knowledge base.

An illustrative example of an SPC-FAPC strategy, published in the Proceedings of the Institution of Mechanical Engineers, Part B, is discussed below.

EXAMPLE 9.1. A process comprising the repetitive positioning and spot welding of a cup (C) onto a sheet metal channel (B), as shown in Figure Ex 9.1. The cup is picked up from the storage bin and positioned over the channel by a hydraulically driven actuator arm (A). The critical dimension is D, which has a target value of 5 mm. Small adjustments to the value of D are governed by FAPC through a control valve. Larger adjustments to the cup position are through the action of SPC guided maintenance intervention.

Typical observations of the uncontrolled rolling average and range of dimension D are shown in Figures Ex 9.2(a) and (b). The sample lots are of five items each.

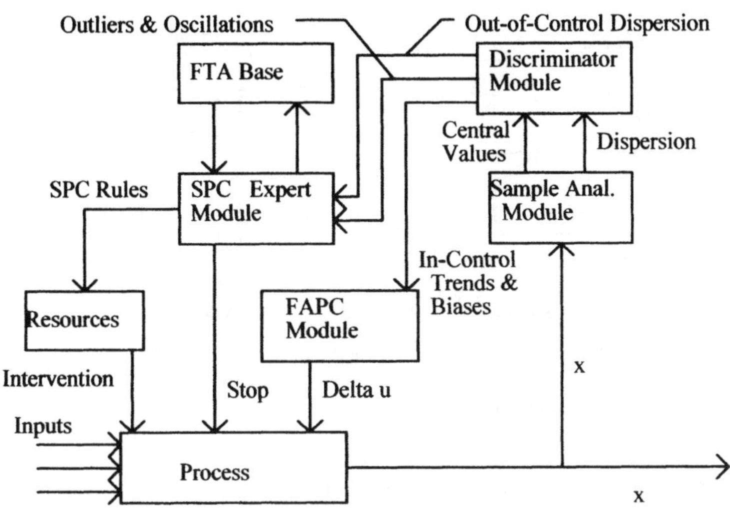

Figure 9.6. SPC-FAPC control strategy control block diagram.

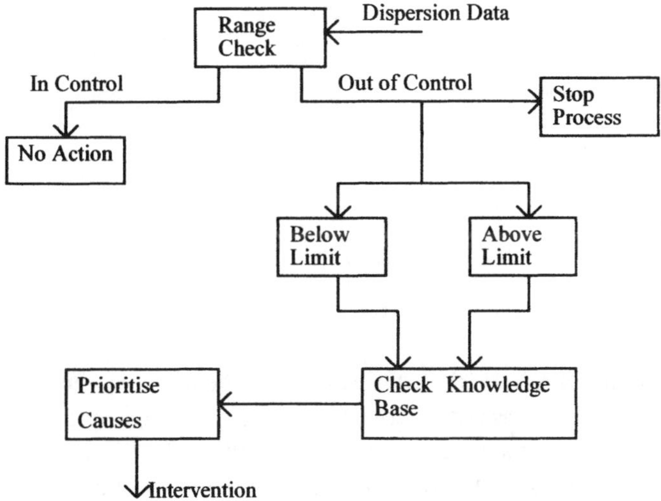

Figure 9.7. Expert module functions for dispersion data.

UCL and LCL of the central values are set at 6 and 4 mm respectively, whilst those of range are at 0.32 and 0.05 respectively. FAPC governs the central values of the UCL-LCL (the in-control) zone, whilst SPC governs the external (out-of-control) zone. The specification of D is 5 mm $\pm$ 10%.

Find the control action required.

Solution. From Figure Ex 9.2(b) it is clear that all the range values are within limits; no control action is required (only SPC is associated with range control). In

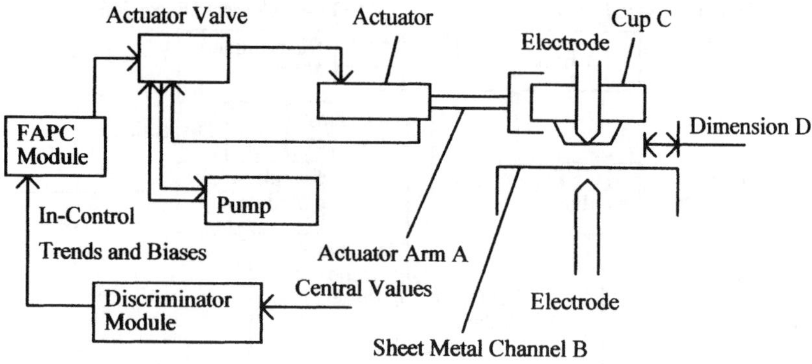

Figure Ex 9.1. Cup to channel spot welding process.

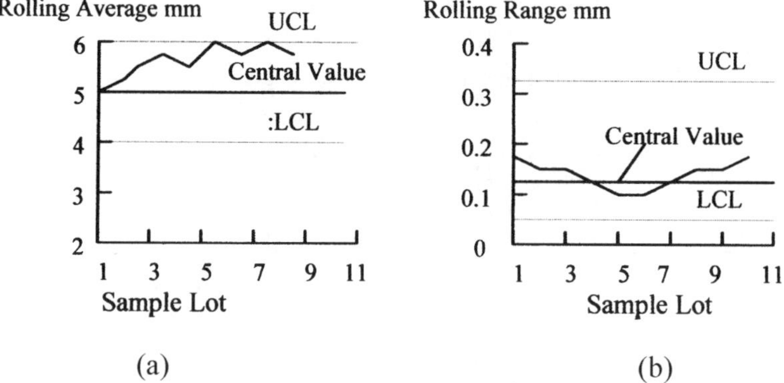

Figure Ex 9.2. Observations of rolling average and rolling mean (uncontrolled). (a) Rolling average. (b) Rolling mean.

the case of central values, Figure Ex 9.2(a), FAPC would be initiated after lot 2. The current outcome of dynamic interaction between the controls and the process would depend upon the ratio of the process time-scale to that of the FAPC action and the sampling time-scale. It may be noted that the effect of using rolling averages and ranges is to suppress the effects of random fluctuations and oscillations in the data values.

The relationship between the actuator arm position and the control signal, u, from the FAPC module output will generally be non-linear. For illustration let the process transfer function in linear form be

$$\delta z_n = a \delta u_n, \quad \text{where } a = -0.5.$$

For the present it will be assumed that the partitioning of the z, δz and δu spaces are as shown previously in Figures 9.2–9.4 respectively and also that the PI controller relational base is given by Table 9.1. To illustrate the FAPC action, assume in

Equation (9.5) that $z_0 = 0$, $z_1 = 0.2$ and $\delta z_1 = 0.2$. Membership values of fuzzy sets may be obtained from Figures 9.2–9.4 or from the membership functions, thus,

$$\mu_{QL1} = 0.5; \quad \mu_{QL2} = 0.5; \quad \mu_D = 1.0.$$

The fuzzy logic proposition is expressed as follows:

IF Z	AND ΔZ	THEN U	MINIMUM	CONSEQUENCE
QL1	D	K	0.5, 1.0	0.5K
QL2	D	J	0.5, 1.0	0.5J

The fuzzy conclusion is the union of the consequences, i.e.,

$$\Delta U_1 = 0.5\text{K} \cup 0.5\text{J}.$$

A controller normally delivers the equivalent of a numerical output which in this case may be represented by some defuzzified form of ΔU_1, obtained for example by the weighted membership formula,

$$\delta u_i = \left(\sum \mu_i \delta u_i \right) \Big/ \sum \mu_i,$$

$$\delta u_2 = (0.5 * 0.1 + 0.5 * 0.2)/1.0 = 0.15.$$

From the process transfer function,

$$\delta z_2 = -0.05 * 0.15 = -0.075.$$

Now

$$z_2 = z_1 + \delta z_2 = 0.2 - 0.075 = 0.125.$$

This completes the first cycle. The second cycle is similar, but in this case (and more generally) the proposition has four terms as follows:

The membership values for new z and u are obtained from Figures 9.2–9.4, as before,

$$\mu_{QL1} = 0.687; \quad \mu_{QL2} = 0.313; \quad \mu_A = 0.75; \quad \mu_B = 0.25.$$

The second proposition is

IF Z	AND ΔZ	THEN U	MINIMUM	CONSEQUENCE
QL1	A	L	0.687, 0.75	0.687L
QL1	B	K	0.687, 0.25	0.25K
QL2	A	K	0.313, 0.75	0.313K
QL2	B	K	0.313, 0.25	0.25K

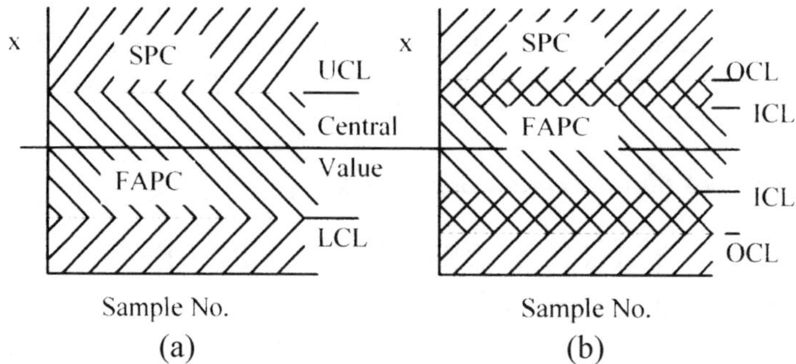

Figure 9.8. Illustration of SPC and FAPC action zones. (a) With UCL/LCL boundaries. (b) With OCL/LCL boundaries.

Then the conclusion is

$$\Delta U_3 = 0.687\text{L} \cup 0.313\text{K}$$

The defuzzified equivalent of the expression is $\delta u_3 = 0.1$. Therefore,

$$\delta z_3 = -0.05 * 0.1 = -0.05,$$

$$z_3 = z_2 + \delta z_{3i} = 0.125 - 0.05 = 0.075.$$

This lies within the prescribed limits of ± 0.10.

SPC

According to the definitions and strategy followed in this work, the SPC protocol will be activated above and below the UCL and the LCL respectively, as shown in Figure 9.8(a). If, however the alternative OCLs and ICLs are defined, then the protocol will be activated as shown in Figure 9.8(b). In the latter case there is a soft boundary between the FAPC and SPC action zones which permits a graded engagement of the external intervention under the SPC regime as the FAPC approaches the limit of its capabilities. This can be formalised by introducing fuzzy sets to represent the FAPC and SPC zones.

SPC action is initiated by the diagnosis of sample average outliers and oscillations in the discriminator module, Figure 9.6 and by out-of-control dispersion data. Then by consultation with its library of causal rules for both sample averages and ranges, the expert module is able to advise and prioritise remedial action. The rules may be deduced from FTAs for each type of out-of-control event, such as actuator position error. FTAs are usually created during the process design stage. A simplified FTA for the actuator in Figure Ex 9.1 is illustrated in Figure 9.9.

The failure probabilities due to the prime faults are shown in Figure 9.9 and these enable the ordering of the most likely causes of error in the actuator position. The prime faults are shown as circles. Each of these could be taken in turn as

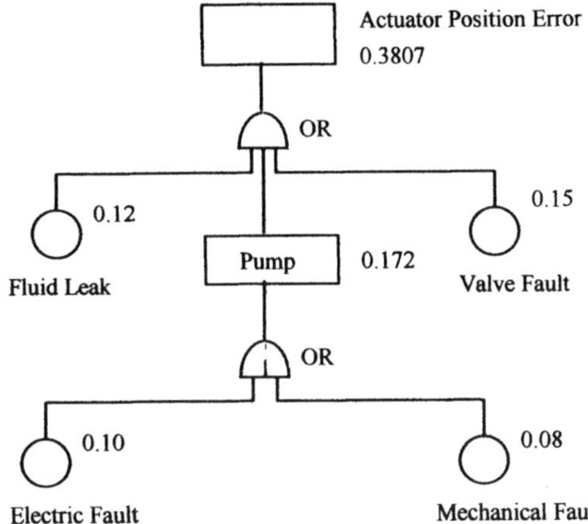

Figure 9.9. Simplified fault tree for the actuator position error (top event).

top events for a fault tree, taking the analysis to greater levels of detail. From Figure 9.9 the likelihood of position error may have the following causes, in order of probabilities:

(i)   Valve fault   0.15.
(ii) Fluid leak   0.12.
(iii) Electrical fault   0.10.
(iv) Mechanical fault   0.08.

The above list also gives the preferred order of investigation for corrective maintenance. The dispersion causes are usually more fundamental, as mentioned earlier.

9.6. Remarks

SPC and FAPC are complementary control functions that can be integrated together to form a comprehensive control strategy. The broad brush portrayal of a transfer function in fuzzy logic terms as a relational array is effective and much simpler than alternative mathematical forms in achieving the desired result. Because no mathematical expression is used, no specific assumption is made whether the controller is either linear or non-linear. Changes in the controller response can easily be made by adjusting the sequence of symbols in the transfer function relational array and the controller can also be made self-adaptive.

Only the control function is portrayed here in FL terms, but the process could also be similarly described if its behaviour is uncertain or complicated.

Chapter 10
Total Risk and Reliability with Human Factors

Keywords. Risk, Reliability, Human Factors, Value-at-Risk, Fuzzy Logic, Fault Tree.

Abstract. Risk is described in terms of an objective and a subjective part. The objective part is expressed as the probability of an unfavourable event, whilst the subjective part is expressed as value-at-risk (VaR). In this work a locomotive power train with track-side and on-board warning signals is considered. The supervisory human element is included in the system and its failure probability is portrayed through the application of fuzzy logic, which is more general than mathematical modelling. This enables human factors to be included in the fault tree analysis (FTA), yielding a comprehensive system failure estimate. The system failure trend is found for a range of human capacities. The financial liability of an organisation is represented by the product of the system failure probability and the VaR.

10.1. Introduction

Risk is becoming an increasingly serious issue in the recent past in all areas of society and has moved up the agenda in corporate thinking to such an extent that it has been argued by Butterworth [1] that the Risk Manager should have Board of Director status, reflecting its importance and possible impact on the reputation and viability of an organisation, especially in view of the increasingly litigious nature of society. The Risk Manager will assess not only the likelihood of a physical incident, but also the immediate costs of system or product unreliability or unavailability, and also the longer term penalties. The total financial consequences are summarised as value-at-risk (VaR), Glasserman [2]. The VaR increases in meaning if it is scaled against nett asset value of the organisation for a commercial organisation, or scaled against the cost centre annual income if it is non-commercial, such as a military or a governmental organisation.

The concept of risk generally implies an objective quantitative estimate of the probability of an event measured by its magnitude or frequency. In industrial systems, unreliability (or unavailability) is usually identified with the objective element in risk. The consequences of risk, identified above as VaR, are subjective and according to Fenton-O'Creevy and Soane [3], more a matter of opinion. The conjunction of the objective and subjective aspects of risk define the concept of criticality, which is useful in selecting defensive strategy for managing risk consequences. The usual definition of reliability, the complement of unreliability, is that it represents the probability that a component, sub-system or system of hardware (or software) will function within specification for a given period of time. This is often assessed by fault tree analysis (FTA) or failure mode and effect analysis (FMEA) through the application of Boolean logic which is well suited to TRUE or FALSE (0,1) propositions, see for example Davidson [4] or Andrews and Moss [5].

The computation of reliability is strongly dependent upon probability theory, particularly upon approximations to the Newtonian formula for the likelihood of favourable events, Margenan [6]. One such is the Gaussian (normal) distribution function, where the total number of events and favourable events are both very large and the approximation is smoothed to a continuous distribution. The Poisson distribution is another approximation, valid when the number of events is large and the number of favourable events is sufficiently small for the mean of the product to be about the order of unity. Probability is only concerned with random events in a stationary system, either of which condition may not be met with exactitude in a system or process.

The human element is an influential feature in system reliability, but unlike the hardware components and sub-systems, it does not have an easily identifiable deterministic form. Whereas in-service artefacts normally exhibit a steadily deteriorating and increasingly unreliable condition over a period of time, the human element performance may be approximately constant in the transport industry, because the operational personnel are drawn from a pool of individuals. But during the span of a working shift human performance maybe expected to fade. In this case, the statistics are spread, not only over the available service population, but also over each daily working shift as individual fatigue arises. This situation is not mathematically modelled in the present work, but made the subject of fuzzy logic (FL) analysis. The human element is important and needs to incorporated into the total reliability assessment; loss of system function (or product function) due to unreliability is a commercial matter, but unreliability causing a risk to health, safety or the environment could be a criminal offence, see for example, Wong [7]. Transport systems are particularly exposed to the whole spectrum of risk consequences.

10.2. The Structure of Total Risk

It is generally accepted that the concept of risk comprises an objective part, which is treated by the examination of historic data and the application of statistics, where appropriate. This is concerned with the uncertainty of events. The other part is subjective , concerned with value judgements of the effects of events, generally called VaR, which is the complement of the more common concept of utility. Figure 10.1 illustrates the components. This work is concerned with the objective aspects of risk, outlined within the overall context of risk below. The subjective part is indicated for completeness of presentation.

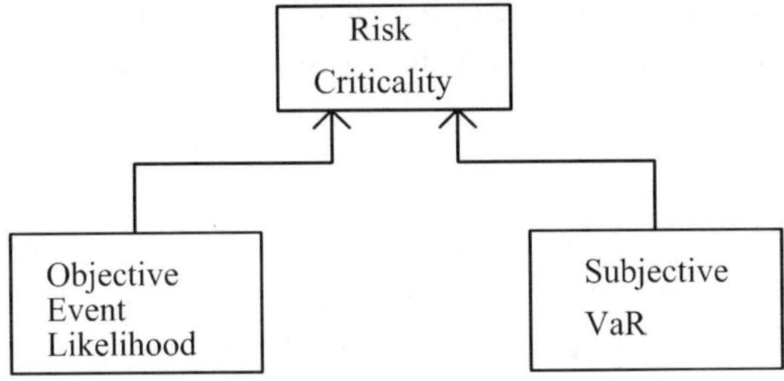

Figure 10.1. The basic structure of risk.

Figure 10.2. Categories of risk.

It is customary to identify unreliability, with the objective side of risk. To place the present work within the wider context it is useful to briefly consider the general field of risk. According to Knight and Pretty [8], there are four categories of risk, as illustrated in Figure 10.2. The following list of categories is adapted from their work. These categories are:

(i) Hazard risk. This relates to physical damage, health and safety, environmental and similar damage which can be the source of liabilities. Also to commercial interruptions, industrial espionage, fire, lightening strikes and data losses.

(ii) Operational risk. Worker unavailability, maintenance stoppages, technical problems, out-of-specification products, business problems, hardware or software failure and fading system performance.

(iii) Commercial risk. Deteriorating market conditions, intelligence reports and information failure or leakage, product obsolescence, hardware obsolescence, exchange rate deterioration.

(iv) Strategic risk. Management decisions on strategic matters and extreme hazards, including force majeure.

Each of the four categories may contain both an objective and a subjective risk component. The human element is a significant factor in categories (i) and (ii).

The main concern in the present work is with the objective aspect of hazard and operational risk. For completeness, a note on VaR is given in the Appendix.

10.3. Human Factors Analysis

In the well-known top-down FTA approach to system failure, a failure mode is postulated and propagated down the fault tree to those components and sub-systems which could have contributed to that specific failure mode. At a particular service time, the failure probabilities are assigned and the resulting system unreliability is found by the application of Boolean logic, see for example, Davidson [4] or Andrews and Moss [5]. In the conventional FTA, the human element remains unaccounted for, although it can have a significant influence. Unlike the system hardware, the human element does not have a convenient mathematical model of the form,

$$R = \exp -t/\lambda, \tag{10.1}$$

see for example Davidson [4] or Harris [9].

A treatment is given here in which the human element is portrayed in terms of fuzzy logic (FL). The human factors arising from the FL analysis are incorporated into the analysis, yielding a total FTA for the whole system, including the operator. In the FL methodology outlined below, it is postulated that the human element observes instrument warning signals both from inside the system (the locomotive drive system) and external from the trackside. The observations and the associated human conditioned reactions together constitute the human element contribution to the FTA. Causes of the alertness of the observations and the nature of the reactions to them are the subjects of psychology theory, and will not be considered further in this work. The key factor here is the probability of a timely and effective response given an internal or an external warning signal, depending on the state of the element.

The functional capacity of the human element is governed by its state at any point in time, and its share between the internal and external signals. Let the capacity of the human element be assigned on a 0–1.0 scale, where 0 means nil capacity for understanding and reacting to warning signals, such as that displayed by an untrained operator, and 1.0 represents the a maximum capacity normally expected of a fully trained, alert and motivated operator. Most system operators would be rated between these two extremes, with a value that would tend to fade from the beginning to the end of a working period.

The human capacity scale is partitioned into a number of overlapping fuzzy sets, as shown in Figure 10.3. The number of fuzzy sets, that is the level of partitioning of the space, called the universe of discourse,(as in statistics), reflects the certainty of their definitions.

The partitioning in Figure 10.3 results in human capacity grades which are present in more than one set, this is a feature of fuzzy sets. The set membership

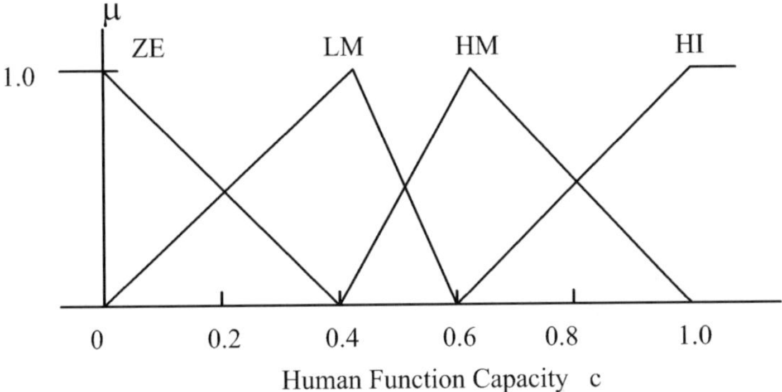

Figure 10.3. Partitioning of the human function capacity. ZE = Zero. LM = Low Medium. HM = High Medium. HI = High.

Table 10.1. Human capacity function membership functions.

	$0 \leq c \leq 0.4$	$0.4 \leq c \leq 0.6$	$0.6 \leq c \leq 1.0$
μ_{ZE}	$1 - c/0.4$		
μ_{L}	$c/0.4$	$3 - c/0.2$	
μ_{HM}		$(c - 0.4)/0.2$	$2.5 - c/0.4$
μ_{HI}			$(c - 0.6)/0.4$

values of a specific grade is given by the membership functions as in Table 10.1. For example, a human factor grade of 0.8 yields membership values of 0.5 in set HM and HI.

The human failure probability universe of discourse is likewise partitioned and an example is shown in Figure 10.4 of equi-partitioning into five fuzzy sets. Table 10.2 shows the corresponding membership functions.

Table 10.2 shows the membership functions for the partitioning illustrated in Figure 10.4.

Table 10.2. Failure probability membership functions.

	$0 \leq p \leq 0.1$	$0.1 \leq p \leq 0.2$	$0.2 \leq p \leq 0.3$	$0.3 \leq p \leq 0.4$
μ_{LO}	$1 - p/0.1$			
μ_{LM}	$p/0.1$	$2 - p/0.1$		
μ_{ME}		$(p - 0.1)/0.1$	$3 - p/0.1$	
μ_{HM}			$(p - 0.2)/0.1$	$4 - p/0.1$
μ_{HI}				$(p - 0.3)/0.1$

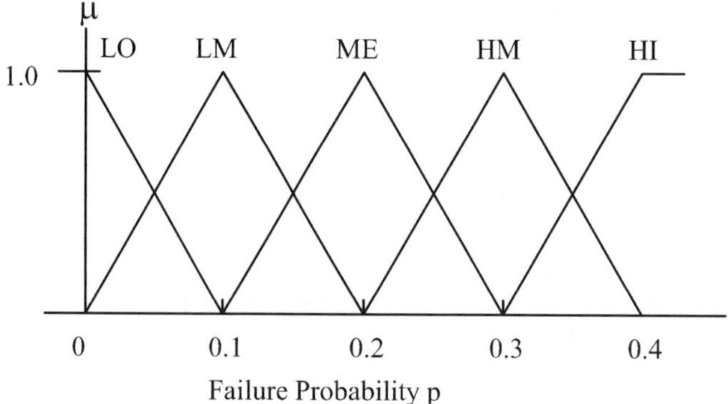

Figure 10.4. Partitioning of the failure probability universe of discourse. Key: LO = Low. LM = Low Medium. ME = Medium. HM = High Medium. HI = High.

Table 10.3. Human capacity and failure probability relational array.

Track-Side	On-board			
	ZE	LM	HM	GD
ZE	HI	HM	ME	LM
LM	HM	ME	LM	LO
HM	ME	ME	LM	LO
GD	ME	LM	LO	LO

The human capacity function and failure probability sets are related by a heuristic array, which is derived from a study of system performance records of similar and equivalent cases to that under consideration. An example of a simple array is given in Table 10.3.

The two-dimensional form of the array in Table 10.3 arises from the capacity of the human element being shared between the on-board (c_b) and the track-side (c_t) warning signals. Thus the total capacity is

$$c_b + c_t = c_u, \tag{10.2}$$

where $0 \leq c_b, c_t \leq 1$. For robot-like perfection, $c_u = 2$. In practice, lack of concentration, diversions or ill-health, for example, result in c_u falling short of this limit.

The probability conclusions result from the intersection operations on the antecedents, Harris [11], in a FL proposition of the form:

IF A   AND B   THEN P

as shown a posteriori.

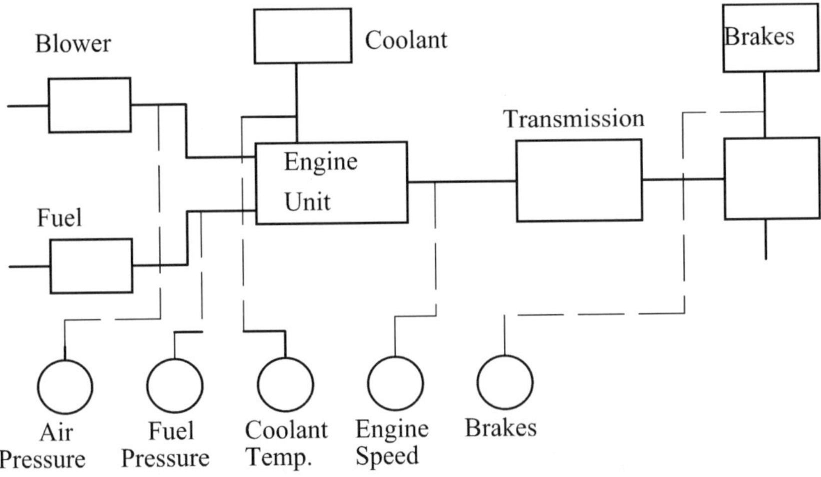

Figure 10.5. Block diagram of locomotive power train.

10.4. Locomotive Power Train

10.4.1. DESCRIPTION OF THE SYSTEM

As an example of the integration of the human element into a FTA, consideration is now given to the traction system of a diesel locomotive. The system considered here comprises:

(i)   Engine unit with blower, fuel supply and cooling water ancillary systems,
(ii) Transmission sub-system,
(iii) Brake sub-system.

The block diagram of the system is shown in Figure 10.5 and includes the on-board warning signals presented to the driver.

The logic diagram of the power train system illustrated in Figure 10.5 is shown in Figure 10.6. For each sub-system, the prime events which could cause failure are: the equipment hardware, the warning instrumentation or the human element, these are labelled in Figure 10.6, and the failure probability at a certain service time is tabulated in the figure. The three prime events are linked together by an inclusive OR operation in each case.

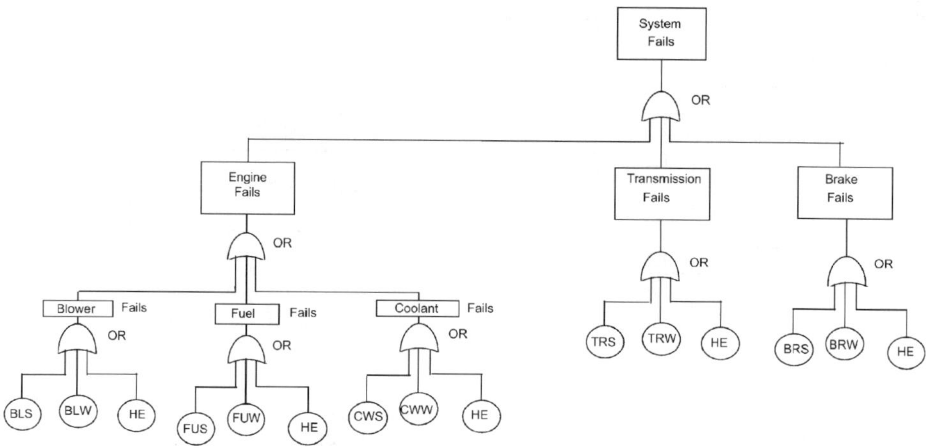

Figure 10.6. Reliability logic diagram for locomotive power train example.

Key:	Unit No.	Code	Type	Failure probability
	1	BLS	Blower Unit	0.043
	2	BLW	Blower Air Pressure	0.022
	3	FUS	Fuel Supply	0.052
	4	FUW	Fuel Supply Pressure	0.022
	5	CWS	Coolant System	0.031
	6	CWW	Coolant Temperature	0.022
	7	BRS	Brake System	0.047
	8	BRW	Brake Warning	0.022

In all the above units in Figure 10.6, HE denotes Human Element. The probability of failure of this element is required to complete the information for analysis.

10.4.2. FAULT TREE ANALYSIS

An FTA can be conducted on the system with the inclusion of the human element failure probability. Consider for example a case in which the capability is shared unequally, with an on-board grade of $c_b = 0.7$ and a track-side grade of $c_t = 0.8$. From Table 10.1, the following fuzzy set membership values may be calculated:

$$\mu_{\text{HM}} = 2.5 - c_b/0.4 = 0.75, \quad \mu_{\text{HM}} = 2.5 - c_t/0.4 = 0.5,$$

$$\mu_{\text{GD}} = (c_b - 0.6)/0.4 = 0.25, \quad \mu_{\text{GD}} = (c_t - 0.6)/0.4 = 0.5.$$

The FL proposition is

IF A	AND B	THEN P	MIN	CONSEQUENCE
HM	HM	LM	0.75, 0.75	0.75LM
HM	GD	LO	0.75, 0.5	0.5LO
GD	HM	LO	0.25, 0.5	0.25LO
GD	GD	LO	0.25, 0.5	0.25LO

The conclusion is the union of the consequences:

$$P = 0.75\text{LM} \cup 0.5\text{LO}. \tag{10.3}$$

This is the human element failure probability in fuzzy set form. To perform a FTA a numerical value is required, which may be found by defuzzifying Equation (10.3) (with some loss of information). A defuzzifying formula is given by (see for example Harris [10])

$$p_{\text{HE}} = \sum \mu_i p_i \Big/ \sum \mu_i \quad (i = 1, 2, \ldots). \tag{10.4}$$

Thus,

$$p_{\text{HE}} = (0.75 * 0.1 + 0.5 * 0)/1.25.$$

Hence, $p_{\text{HE}} = 0.06$.

This failure probability of the human element becomes a prime cause and applies to each unit, as illustrated in Figure 10.6. The resulting FTA proceeds in the normal way.

Considering each unit separately, and referring to Figure 10.6:

Blower.

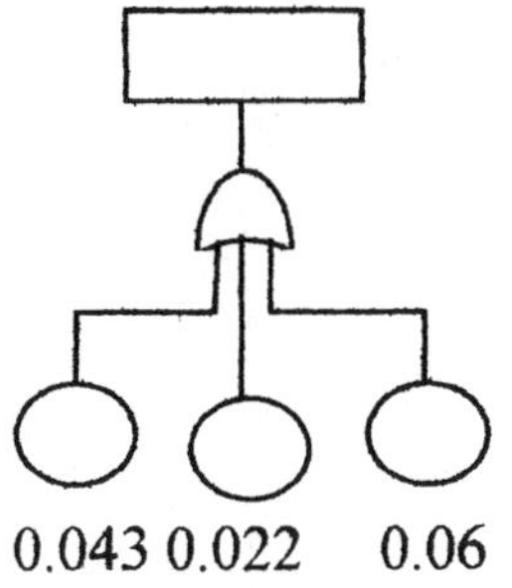

The success probability (q) of the blower unit is the intersection of the complements of the failure probabilities,

$$q = (1 - 0.043)(1 - 0.022)(1 - 0.06) = 0.8793.$$

Hence, $p = 1 - q = 1 - 0.8793 = 0.1202.$

Fuel System.

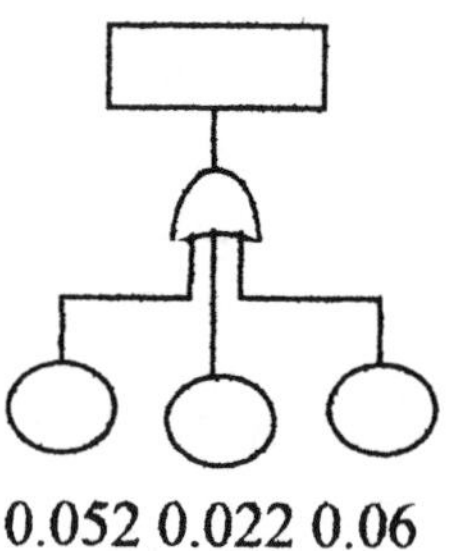

Similarly, the success probability if the fuel system unit is

$$q = (1 - 0.052)(1 - 0.022)(1 - 0.06) = 0.8715,$$

$$p = 1 - 0.8715 = 0.1285.$$

The success and failure probabilities of the remaining units are as follows:

Cooling Unit.

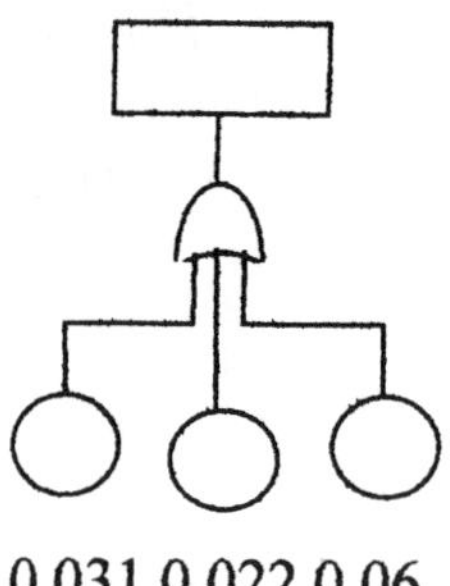

$$q = 0.8908; \quad p = 0.1092.$$

Engine Unit.

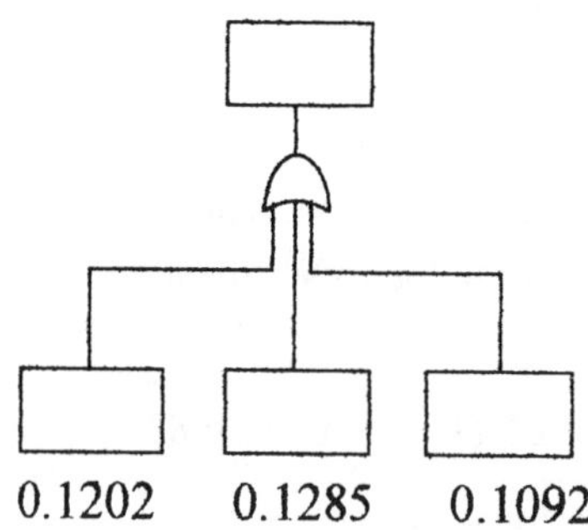

$$q = 0.6830; \quad p = 0.3170.$$

Transmission Unit.

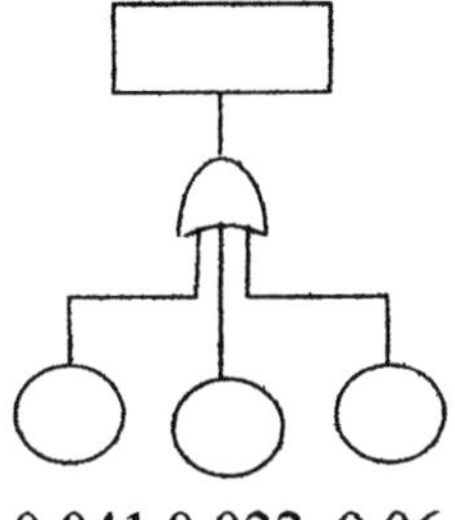

$$q = 0.8816; \quad p = 0.1184.$$

Brake Unit.

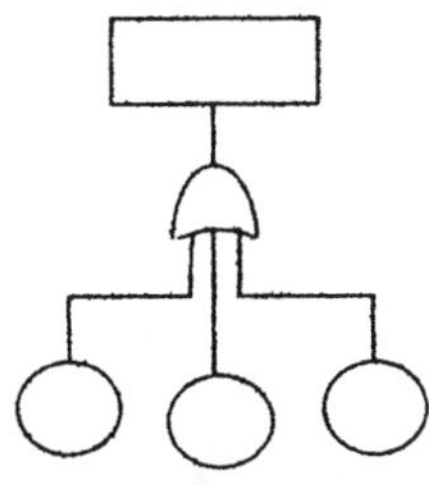

$$q = 0.8716; \quad p = 0.1239.$$

Power Train System.

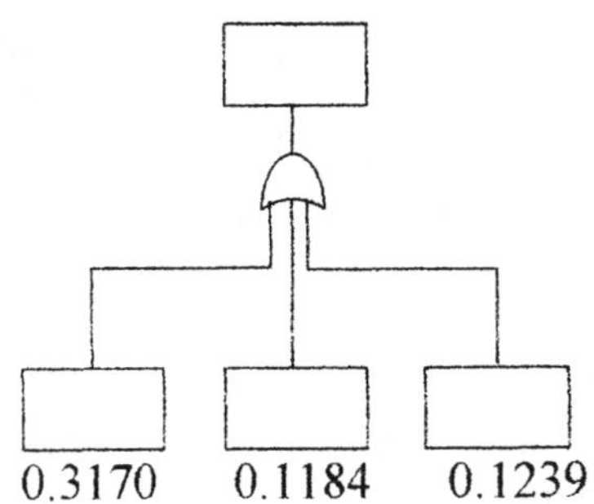

$$q = 0.8908; \quad p = 0.1092.$$

where p_{SY} is the overall failure probability of the power train system.

To illustrate the trends with varying values of c_b and c_t the values of p_{HE} and p_{SY} been calculated for several equal values of c_b and c_t. These are given in Table 10.4 and displayed in Figure 10.7.

10.5. Criticality

The VaR (outlined in the Appendix) provides a measure of the potential financial liabilities of an organisation arising from an unfavourable event, which requires

Table 10.4. Values of p_{HE} and p_{SY} for $c_b = c_t$.

c_b (= c_t)	p_{HE}	p_{SY}
0	0.4 (min.)	0.9441
0.1	0.34	0.9652
0.2	0.25	0.8294
0.3	0.26	0.9652
0.4	0.20	0.6923
0.5	0.15	0.6810
0.7	0.075	0.5132
0.8	0.05	0.4446
0.9	0.025	0.3666
1.0	0	0.2078

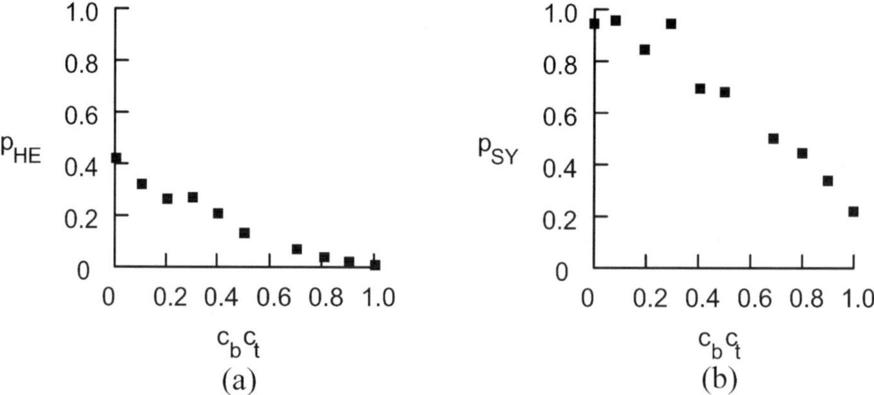

Figure 10.7. Illustration of (a) the Human Element and (b) the System Failure Probability Trends.

defensive management action. This would often be sought through insurance cover requiring a premium (x) for a selected duration. If in this period the system failure probability is p_{SY}, the criticality of the case (y) is defined by

$$y = \text{VaR} * p_{SY}. \tag{10.5}$$

If $x > y$ it may not be worth seeking external insurance cover. Financial considerations are not further developed in this work.

The system failure probability represents the expected frequency of unfavourable events (in the long term). Therefore, for a given service duration of similar systems under similar service conditions, the number of unfavourable events can be calculated and hence the possible corporate financial liability, or criticality, of this type of event can be found. Extreme events are usually characterised by their magnitude rather than by their frequency, they are in another class and are not included in this work.

10.6. Conclusion

In this work, a comprehensive estimate of the system failure probability is found by integrating the human element effects into the analysis. The human element failure probability does not share the same kind of time-dependency as the system hardware. The capacity (or capability) of the human element to perform its ideal supervisory and governing role is shared between track-side and on-board warning signals in the case of a locomotive driver function. The proportional share would depend upon the prevailing environmental conditions and the total human capacity available, the latter would depend upon the person's state. It would naturally be expected, for example, that the state of alertness would decline during a working period.

The human element role is portrayed in terms of fuzzy logic, which avoids the need for mathematical modelling, and is much more general in scope. The heuristic relational base can be readily updated as operational experience is gained. Relatively coarse partitioning of the universes of discourse are used, which may also be refined as knowledge and experience of the human function increases.

The probability of system failure provides the objective side of risk. The potential financial damage to an organisation is expressed in terms of the VaR, which seeks to quantify, in financial terms, the subjective side of risk. The probability of a certain level of financial liability at a given point in time is expressed as the product of the failure probability with the VaR.

10.7. Notation

μ_A, μ_B	membership values of fuzzy sets
λ	time constant
$\cup$	logic union

p_{SY}	total system failure probability
p_{HE}	human element failure probability
c_b	on-board human element capacity
c_t	track-side human element capacity
c_u	total human element capacity $= c_b + c_t$
x	insurance cover premium
y	system failure criticality, defined by Equation (10.5)

$P()$	probability of failure (objective risk)
R	reliability

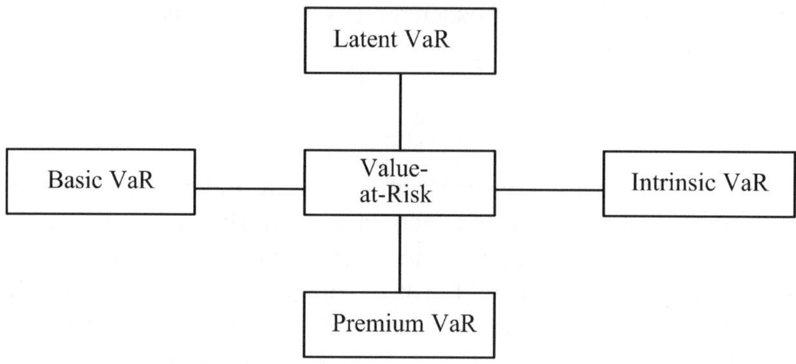

Figure 10.A1. Composition of the value-at-risk.

Appendix. Comments on Value-at-Risk

The subjective side of risk, expressed by the VaR, has been quite widely considered recently, mainly in economics and finance, see for example Glasserman [2], Fenton-O'Creevy and Soame [3], Carey and Turnbull [11] and also Lewin [12]. VaR is also a composite concept comprising four components as illustrated in Figure 10.A1. The components are:

(i)   *Basic value.* This represents the accounting value or book value of an asset at risk, taking into consideration depreciation and other similar factors.

(ii) *Premium value.* Additional value of the service provided by an asset. Health and safety liabilities.

(iii) *Latent value.* Enterprise reputation. Potential developmental value of an asset at risk. Political credit.

(iv) *Intrinsic value.* Value due to an historic or cultural interest.

These values are to be aggregated and scaled against a basic measure such as the nett worth of an enterprise, or the gross annual income of a non-commercial enterprise.

In the case treated in this work, it is the latent, basic and premium categories that are likely to be relevant.

References

1.   Butterworth, M., "The Emerging Role of the Risk Manager", *Mastering Risk*, Part 1, Financial Times Supplement (UK), April 25, 2000, pp. 12–13.

2.   Glasserman, P., "The Quest for Precision through Value-at-Risk", ibid., Part 4, May 16, 2000, pp. 6–7.

3.   Fenton-O'Creevy, M. and Soane, E., "The Subjective Perception of Risk", ibid., Part 1, April 25, 2000, pp. 14–15.

4. Davidson, J. (ed.), *The Reliability of Mechanical Systems*, 2nd edn, Mech. Engng. Publ. (UK), 1994, ISBN 0-85298-881-8.

5. Andrews, J.D. and Moss, T.R., *Reliability and Risk*, Longman Scientific and Technical (UK), 1993, ISBN 0-582-09615-4.

6. Margenau, H., "Probability (Physics)", in *McGraw-Hill Encyclopedia of Science and Technology*, 19, Vol. 5, pp. 401–404.

7. Wong, W. (ed.), *Process Machinery-Safety and Reliability*, Mech. Engng. Publ. (UK), 1997, ISBN 1-86058-046-7.

8. Knight, R. and Pretty, D., "Philosophies of Risk, Shareholder Value and the CEO", ibid., Part 10, June 27, 2000, pp. 14–15.

9. Harris, J., "Piecewise Linear Reliability Data Analysis with Fuzzy Sets", *Proc. Instn. Mech. Engrs.*, Vol. 215, Part C, pp. 1075–1082.

10. Harris, J., *An Introduction to Fuzzy Logic Applications*, Kluwer Academic Publishers, Dordrecht, 2000.

11. Carey, A. and Turnbull, N., "The Boardroom Imperative on Internal Control", ibid., Part 1, April 25, 2000, pp. 6–7.

12. Lewin, C., "Refining the Art of the Probable", ibid., Part 2, pp. 2–4, May 2, 2000.

Chapter 11
On System Condition Auditing [*]

Keywords. Condition metric, Fuzzy logic, Risk, FTA, Maintenance, Knowledge management.

Abstract. Condition monitoring of a variety of systems is widely practised and yields valuable information about the current status of key functional parts of the system. This is normally used for the organisation of maintenance functions of industrial systems (or for predictive treatment in the case of health care). At the executive level, asset management is an important consideration, and is usually based on financial grounds alone. In this work the methodology of system condition auditing based on fuzzy logic analysis is described, which results in an overall system condition metric. This makes fuller use of the information garnered from condition monitoring and provides additional management knowledge, enabling more informed decisions to be made about such aspects as plant refurbishment or replacement.

11.1. Introduction

The term "system" in this context connotes any coherent artificial or natural assembly of parts which may perform static or dynamic functions. Typical examples of various classes are: structures and buildings, process plant, surface or airborne vehicles and also biological systems. Though for illustration purposes, the discussion in Section 3 will focus on process plant. Condition monitoring is widely practised in all sectors of engineering and is at its most intense and sophisticated in the aero-space sector. It also is very refined in the nuclear industry and in health care in the more industrialised countries.

At the design stage of artificial systems, reliability analysis and also material sample, model, prototype and product testing are often undertaken to compare with performance specifications and as a guide for future maintenance strategies. This is augmented by in-service condition monitoring. With large-scale one-of-a-kind systems, such as civil engineering structures, the initial testing is restricted to material samples and system models. Reliability analysis is itself therefore of more limited value, due to a number of factors listed below, similar to those given by Geeker [1]:

---

[*] This material has been reproduced from *Trans IChemE, Part B, Process Safely and Environmental Protection*, 2002, Vol. 80, pp. 197–203, "On System Condition Auditing" by J. Harris, with permission of the Council of the Institution of Chemical Engineers.

  (i) inadequate system deterioration models,
 (ii) inadequate load and environmental models,
(iii) inaccurate testing and analysis methods,
(iv) insufficient information.

Fatigue testing has been one of the commoner methods of providing basic reliability data (see Chapter 6). Even if the factors in the above list did not exist, there may still be differences between the idealisations of a particular system design and its realisation in practice. Thus, repeated condition monitoring provides updated and improved knowledge of the actual system states and of its deterioration phenomena. It also provides the means of updating reliability and durability forecasts. This facilitates reduced life-cycle costs through improved preventative and corrective maintenance schedules.

Considering that some systems such as bridges, may have a design life of about 100 years, it is clear that predictions of loading and environmental conditions over the life-span can be highly speculative and subject to significant error in the forecasts of future system reliability. At the operational level, system condition observations are compared with target values, which may be qualitative or quantitative (for example, the notion of a building's dilapidation may be partly qualitative). If the system is non-repairable, as in some space satellites, non-conformance to target values may indicate the end of useful service life, whilst in the case of a repairable system, some form of maintenance may alternatively be indicated. Another option in both cases is partial decommissioning by reducing the system's service or environmental demands.

System reliability analysis normally distinguishes only between the failed or non-failed states, which is the lowest level of classification. All systems undergo inexorable deterioration or ageing in time through a variety of agencies such as corrosion, wear or fatigue damage, which may or may not influence performance or value. (Value is not always determined entirely by performance.) Furthermore, a system may perform adequately, but exhibit some deterioration. Such a state would be detected, tracked and measured within an efficient condition monitoring programme.

System condition monitoring is usually undertaken with the support of a variety of non-destructive test (NDT) methods, such as thermal and acoustic emission, ultra-sonics, radiography and fluorescent tests and also by operator visual and audio examination. For large systems, destructive material sample and component tests are sometimes performed. In the high-volume manufacturing industry statistical process control (SPC) is also used as another condition monitoring tool, Wang and Rowlands [5]. There are often Standards or industry based Codes of Practice to guide and regulate the acquisition of condition data. Harris [6] has shown how such Standards may be recast into fuzzy set form to improve their interpretation, using the ultra-sonics British Standard BS 6208:1990 as an example.

In this work, attention is directed towards knowledge management of physical data derived from condition monitoring observations. Not only does this provide a useful comprehensive view of the system state, but also by repeated observations the progress of the gross system condition pattern over time is revealed. One of the benefits of this is that updated and improved system reliability forecasts may be made based upon new knowledge of actual system behaviour.

At management level, comprehensive and comprehensible information is required for making strategic decisions. The resulting methodology described in this work provides a system condition metric which would normally be conjoined at management level with a utility factor (more generally a value-at risk VaR, Glasserman [7], which includes financial as well as safety, environmental, social and other factors deemed by management to be part of the total liability burden of operations).

11.2. Basic Considerations

It is assumed that the system, of whatever type, can be divided into a finite number of functionally independent units which are monitored at the same point in "time" ("time" being any convenient metric of the system condition). The system parameters to be measured should be chosen to be sufficient to define its condition for the purpose of the system audit. (In some cases this may include aesthetic as well as technical features.) The type of observation may be quantitative or qualitative, but in either case it is reported on a 0–1.0 scale

Some similarities may be drawn with the well-known methods of fault tree analysis (FTA), which are widely reported in the engineering literature, see for example Davidson [9], also Andrews and Moss [10]. There are also some significant differences arising from the different purposes of condition monitoring compared with FTA. This means, for example, that the systems considered may not have the same definition in general or in detail. Also FTA is concerned with fail/safe conditions, whereas in the analysis that follows, it is the state of a system's deterioration which is of concern.

11.2.1. BOUNDARY DEFINITIONS

External boundary: This depends upon the purpose of the audit. Systems may rely on supplies from external services which may not be included in the system definition, access may be another important feature. If, for example, the condition audit is for purposes of renovation, then different considerations may apply to those for market valuation.

Internal boundaries: These define the relevant sub-systems or units considered to comprise the system. Units may be taken to the level of components, depending on the level of detail desired and also the information that is available.

In general terms, the external boundary of a system limits the breadth of a condition audit, whereas the internal boundaries define its depth of detail.

11.2.2. MINIMAL PATH SETS

In FTA the application to a particular case results in an estimated probability that a system will fail under specified environmental and service conditions, given the failure probabilities of the basic events. System failure modes are governed by the so-called minimum cut sets, see for example Andrews and Moss [10]. The converse concept is that of path set, which defines a set of basic events which, providing that they occur, ensures that the system will function, in particular, the minimal path set provides the irreducible set for success. For condition auditing, the minimal path set provides the simplest combination of conditions that define a minimum audit for a system. It will be recalled that a condition audit may require more than just consideration of a minimum set of operational units.

11.2.3. MALFUNCTIONS

In this work, the failure of a unit or a component is defined as a malfunction that requires only adjustment by plant operators for correction. It does not affect the system condition audit. On the other hand, a unit or component fault is a malfunction that requires intervention by maintenance staff repair or replacement action. This does affect the system condition audit.

An example of the methodology of system auditing is given below. In practice FTAs may be applied to identify the sources of system failure at which to make system condition observations.

11.2.4. HYPOTHESIS

The underlying hypothesis in this work is that the overall condition of a system may be represented by the logical intersection of the several sub-system conditions.

11.3. Heat Exchanger System

As a simple example of a typical industrial system condition audit, consideration is given in Section 11.4 to a gas/liquid heat exchanger. The external boundary is taken here to encompass the liquid side only to show auditing methodology, and to avoid duplication.

11.3.1. SYSTEM DESCRIPTION

The heat exchanger is of the shell and tube type for cooling process gas with water, in which the gas flows on the shell side whilst the water flows on the tube side.

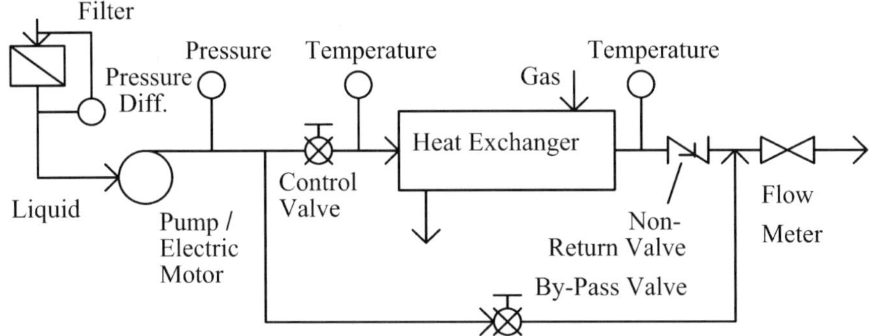

Figure 11.1. Heat exchanger block diagram with liquid side instrumentation.

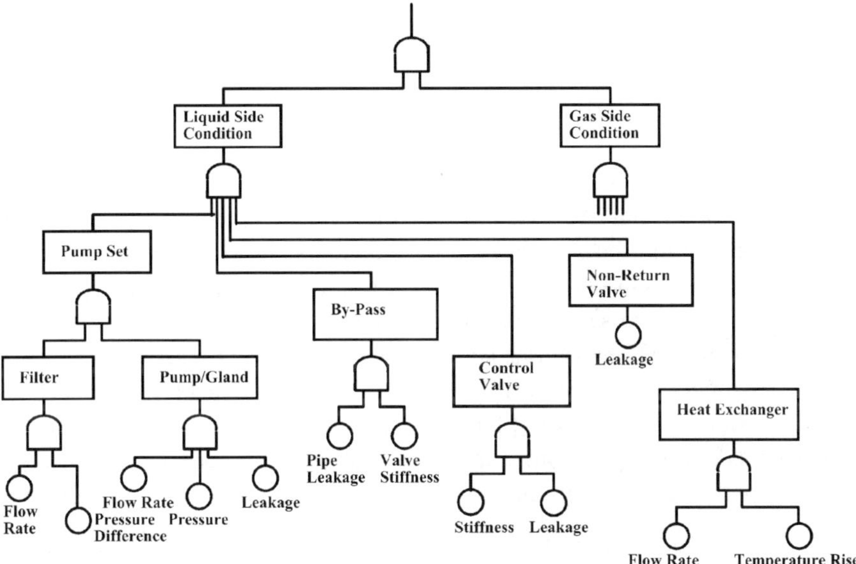

Figure 11.2. Heat exchanger liquid side system condition logic diagram. Prime functions only. Instruments not included.

The system with auxiliaries is illustrated in the block diagram in Figure 11.1. The external boundary includes only the system liquid side in this analysis. System condition data is provided by the instrumentation shown, which provides quantitative data. Some observations are also made on a qualitative basis, including bearing noise, valve and pipe leakage and also valve stiffness. These qualitative observations are also reduced to a 0–1.0 scale for reporting. The minimal path set in this case corresponds with the external boundary.

The electric motor is fixed speed. The pump operating point may, or may not be, at the design operating point, depending upon the state of the system.

Figure 11.2 shows the equivalent condition logic diagram (CLD). This comprises relations between system basic (or prime) events, which are defined in this

Table 11.1. System condition observations.

Unit	Observation	Condition Rating	Design Rating
1. Electric motor	Power	0.6	0.5
	Bearing noise	0.2	0.0
2. Filter/Pump	Flow rate	0.46	0.5
	O/P Pressure	0.56	0.5
	Noise	0.15	0.0
3. By-Pass	Pipe/valve leakage	0.0	0.0
	Valve stiffness	0.15	0.0
4. Control valve	Stiffness	0.0	0.0
	Leakage	0.0	0.0
5. Non-return valve	Shut-off	1.0	1.0
6. Heat exchanger	Flow rate	0.43	0.5
	Temperature rise	0.45	0.5

case as the units that are essential for the system operation. Instrumentation is not included within the system external boundary. On a 0–0.1 scale, the assumed observed prime event conditions are given in Table 11.1. The gates are all of the AND type. Both intersection and logic union operations are used in the analysis, but only intersection operations show on the CLD in this case.

It is assumed that the system in Figure 11.1 and Table 11.1 has been under the of supervision of a competent plant operator and also subject to plant maintenance procedures. The design ratings are those that would be obtained with a "good as new" properly balanced system.

11.4. System Condition Audit

In the following analysis, the universes of discourse (UDs), x, are equi-partitioned into five fuzzy sets, X_i, $i = 1$ to 5. The corresponding membership functions are of a commonly used form (see Harris [2, 3, 6, 8] and also Wang and Rowlands [5]), they are shown in Table 11.2.

In Table 11.2, x denotes any of the UDs in the following analysis and X denotes any corresponding fuzzy set. Although all observations are reduced to a scale of 0–1.0, the range in Table 11.2 is extended from -0.25 to 1.25 to allow completion of the extreme fuzzy sets.

11.4.1. FILTER

The condition observations are: pressure drop and flow rate. The usual deterioration is due to progressive blockage of the filter by debris, giving an increased pressure

Table 11.2. Membership functions (μ) for an equi-partitioned UD.

	$-0.25 \le x \le 0$	$0 \le x \le 0.25$	$0.25 \le x \le 0.5$	$0.5 \le x \le 0.75$	$0.75 \le x \le 1.0$	$1.0 \le x \le 1.25$
X_1	$1+4x$	$1-4x$				
X_2		$4x$	$2(1-2x)$			
X_3			$4x-1$	$3-4x$		
X_4				$2(2x-1)$	$4(1-x)$	
X_5					$4x-3$	$5-4x$

drop for a given flow rate. For a particular clear filter, let the filter pressure drop (i_p) be related to the flow rate (q) by

$$c_f = \Delta p / q, \tag{11.1}$$

where c_f is a constant.

Let c_f' be the observed in-service value, then a filter condition metric may be defined as

$$f = c_f / c_f'. \tag{11.2}$$

Thus, if $c_f = 0.46$ and $c_f' = 0.56$, then $f = 0.8214$.

11.4.2. PUMP AND MOTOR

The condition observations in this case are: the hydraulic head (h) developed for a given fluid volumetric flow rate (q). The leakage rate at the gland is also noted and expressed on a scale of 0–1.0, where 1.0 represents effective gland sealing and 0 represents poor gland sealing.

It may be shown by dimensional analysis that a dimensionless group (c_h), which characterises the performance of an impeller type pump, may be expressed as

$$c_h = (gh/q)(d/n), \tag{11.3}$$

where d is a characteristic pump size and n is the impeller rotational speed. Hence, for a given pump size and speed the characteristic may be expressed as

$$c_h = \text{constant} * (h/q). \tag{11.4}$$

Let c_h be the pump characteristic under design conditions and c_h' be the actual observed condition, then the pump condition observation may be expressed as

$$i = (2/\pi) \tan^{-1} |c_h/(c_h' - c_h)|. \tag{11.5}$$

In the given system, the pump-gland condition is assumed to be 0.8. The design pump characteristic (c_h) is given in Table 11.1 as 0.5 and the observed value (c_h'), is 0.6, hence the condition metric is obtained from Equation (11.5) as

$$i = (2/\pi) \tan^{-1} |0.5/(0.6 - 0.5)|$$
$$= 0.8743.$$

11.4.3. PUMP AND FILTER CONDITION

In Section 11.4.1 the filter condition observation (f) is found to be 0.8214. From Table 11.2, the fuzzy membership values are

$$\mu_{F4} = 0.7144 \quad \text{and} \quad \mu_{F5} = 0.2856.$$

Hence, the fuzzy form of the filter condition observation is

$$F = 0.7144F_4 \cup 0.2856F_5. \tag{11.6}$$

Similarly, in Section 11.4.2 the pump condition observation (i) is given as 0.8743, hence from Table 11.2 the membership values are found as follows:

$$\text{Membership values:} \quad \mu_{P4} = 0.5028 \quad \text{and} \quad \mu_{P5} = 0.4972.$$

The resulting pump condition observation is expressed as

$$P = 0.5028P_4 \cup 0.4972P_5. \tag{11.7}$$

Also, in Section 11.4.2, the gland condition observation is given as 0.8, whence from Table 11.2:

$$\text{Membership values:} \quad \mu_{G4} = 0.8 \quad \text{and} \quad \mu_{G5} = 0.2,$$

i.e.,

$$G = 0.8G_4 \cup 0.2G_5. \tag{11.8}$$

The pump condition metric is the result of the intersection of expressions (11.7) and (11.8),

> IF P AND G THEN PG

where the relationship between the antecedents and the consequents is governed by the rule-base shown in Table 11.3. (All rule-bases used are simple plausible relationships.)

Hence, for the antecedents in Equations (11.7) and (11.8) the following propositions are formed, relating to the pump and gland:

Table 11.3. Rule-base for the pump and gland.

	P				
G	**P$_1$**	**P$_2$**	**P$_3$**	**P$_4$**	**P$_5$**
G$_1$	PG$_1$	PG$_1$			
G$_2$	PG$_1$	PG$_2$	PG$_2$		
G$_3$		PG$_2$	PG$_3$	PG$_3$	
G$_4$			PG$_3$	PG$_4$	PG$_4$
G$_5$				PG$_4$	PG$_5$

IF P	AND G	THEN PG	MIN	CONCLUSION
P$_4$	G$_4$	PG$_4$	0.5028, 0.8	0.5028PG$_4$
P$_4$	G$_5$	PG$_4$	0.5028, 0.2	0.2PG$_4$
P$_5$	G$_4$	PG$_4$	0.4972, 0.8	0.4972PG$_4$
P$_5$	G$_5$	PG$_5$	0.4972, 0.2	0.2PG$_5$

The overall pump and gland condition metric is found by the union of the above partial conclusions,

$$PG = 0.5028PG_4 \cup 0.2PG_5. \tag{11.9}$$

11.4.4. PUMP UNIT CONDITION

The pump unit includes the pump with gland and also the filter. These are conjoined by an intersection operation to provide a resultant pump unit (PU) condition, thus,

$$PU = PG \cap F. \tag{11.10}$$

The relationship is governed by a rule base such as that shown in Table 11.4.

From Equations (11.6) and (11.9) the resultant pump unit condition metric is therefore,

$$PU = (0.5028PG_4 \cup 0.2PG_5) \cap (0.7144F_4 \cup 0.2856F_5). \tag{11.11}$$

Developing the proposition in Equation (11.11):

IF PG	AND F	THEN	MIN.	CONCLUSION
PG$_4$	F$_4$	PU$_4$	0.5028, 0.7144	0.5028PU$_4$
PG$_4$	F$_5$	PU$_4$	0.5028, 0.2856	0.2856PU$_4$
PG$_5$	F$_4$	PU$_4$	0.2, 0.7144	0.2PU$_4$
PG$_5$	F$_5$	PU$_5$	0.2, 0.2856	0.2PU$_5$

Table 11.4. Rule base for the pump unit condition.

PG			P		
	F_1	F_2	F_3	F_4	F_5
PG_1	PU_1	PU_2			
PG_2	PU_1	PU_2	PU_2		
PG_3		PU_2	PU_3	PU_3	
PG_4			PU_3	PU_4	PU_4
PG_5				PU_4	PU_5

The resultant pump unit fuzzy condition metric is therefore,

$$PU = 0.5028PU_4 \cup 0.2PU_5. \tag{11.12}$$

This expression may be defuzzified, see Harris [8], to yield a numerical represent-ation of the condition metric. Noting from Table 11.2 that the maximum value of the PU_4 set is at 0.75 on the UD and for the PU_5 set it is at 1.0, the defuzzified value (def PU) is obtained by the following calculation:

$$\mathrm{def\,PU} = (0.5028 \times 0.75 + 0.2 \times 1.0)/(0.5028 + 0.2)$$

$$= 0.8211.$$

11.4.5. BY-PASS

The by-pass valve is observed for gland sealing (L) and ease of operation (E). Both these are qualitative observations and scored on a 0–1.0 scale. From Table 11.1, the ease of operation is assessed as 0.85 and the sealing is 1.0. It is assumed that there is no pipework leakage. The resultant by-pass condition (B) is derived from the following logic proposition:

$$B = L \cap E. \tag{11.13}$$

Table 11.5 shows the rule-base for the by-pass condition.

If the observed valve gland seal condition is 1.0, then from Table 11.1 the corresponding membership function value is $\mu_{L5} = 1.0$, and clearly, $L = 1.0L_5$.

The operational ease condition is 0.85 and from Table 11.2 the corresponding membership values are as follows:

Membership values: $\mu_{E1} = 0.4$ and $\mu_{E2} = 0.6$.

The fuzzy expression for the operational ease observation is therefore,

$$E = 0.4E_1 \cup 0.6E_2. \tag{11.14}$$

Table 11.5. Rule-base for the by-pass condition.

		E			
L	E_5	E_4	E_3	E_2	E_1
L_1	B_1	B_1	B_1	B_1	B_1
L_2	B_1	B_2	B_2	B_2	B_2
L_3	B_1	B_2	B_3	B_3	B_3
L_4	B_1	B_2	B_3	B_4	B_4
L_5	B_1	B_2	B_3	B_4	B_5

The intersection of Equation (11.14) set with $L = 1.0L_5$ results in the fuzzy by-pass condition metric,

$$B = 1.0L_5 \cap (0.4E_1 \cup 0.6E_2). \tag{11.15}$$

The antecedents in Equations (11.14) and (11.15) provide the following proposition developments:

IF L	AND E	THEN B	MIN.	CONCLUSION
L_5	E_1	B_5	$1.0, 0.4$	$0.4B_5$
L_5	E_2	B_4	$1.0, 0.6$	$0.6B_4$

The resultant by-pass condition metric is therefore represented in fuzzy form by

$$B = 0.6B_4 \cup 0.4B_5. \tag{11.16}$$

As for the pump unit above, the defuzzified value may be found, in this case it is def $B = 0.85$.

11.4.6. CONTROL VALVE

The unit condition observation is "as good as new", hence the resulting condition metric is

$$D = 1.0D_5. \tag{11.17}$$

Clearly, the defuzzified value is def $D = 1.0$.

11.4.7. NON-RETURN VALVE

The observed condition is assumed to be 0.95, hence from Table 11.2 the membership values are as follows:

$$\text{Membership values:} \quad \mu_{R4} = 0.2 \quad \text{and} \quad \mu_{R5} = 0.8.$$

The resultant condition observation is

$$R = 0.2R_4 \cup 0.8R_5. \tag{11.18}$$

The defuzzified value is def $R = 0.95$, which of course corresponds with the observed condition.

11.4.8. HEAT EXCHANGER UNIT

For a given gas/liquid heat exchanger operating in the turbulent region, the performance may be expressed in terms of the Stanton number (s_t) and the Reynolds number (r_e), where these two dimensionless groups may be expressed as

$$s_t = k_1 \eta / \theta q, \tag{11.19}$$

and

$$r_e = k_2 q. \tag{11.20}$$

In Equations (11.19) and (11.20) η and q are the heat load and fluid flow rate respectively and θ is the mean temperature difference, also k_1 and k_2 are two specific constants. For a given heat load, the design Stanton number is s_t. If the observed Stanton number is s_t', the heat exchanger observation may be expressed as

$$t = (2/\pi) \tan^{-1} |s_t / (s_t' - s_t)|. \tag{11.21}$$

If for the given system in Figure 11.1 the Stanton number ratio, $s_t'/s_t = 0.774$, then from Equation (11.21) the corresponding numerical observation is $t = 0.8585$. Therefore, from Table 11.2 the following membership values may be found:

$$\text{Membership values:} \quad \mu_{T4} = 0.566 \quad \text{and} \quad \mu_{T5} = 0.436.$$

Hence, the heat exchanger fuzzy condition may be expressed as

$$T = 0.566T_4 \cup 0.436T_5. \tag{11.22}$$

The defuzzified value in this case is def $T = 0.8588$, which is consistent with the numerical observation.

Table 11.6. Fuzzy set combinations for W.

Set	PU_4	PU_5	B_4	B_5	R_4	R_5	T_4	T_5	Min	Subscript	Result
μ	0.5028	0.2	0.6	0.4	0.2	0.8	0566	0.436		i	
	*		*		*		*		0.2	4	$0.2W_4$
	*		*		*			*	0.2	4	$0.2W_4$
	*		*			*	*		0.5028	4	$0.5028W_4$
	*		*			*		*	0.2	Only effective if $i = 5$	
	*			*	*		*		0.2	4	$0.2W_4$
	*			*	*			*	0.2	Does not affect outcome	
	*			*		*	*		0.4	Does not affect outcome	
	*			*		*		*	0.4	5	$0.4W_5$

11.4.9. SYSTEM WATER-SIDE CONDITION METRIC

The several fuzzy metrics consistent with the available observations of the units comprising the water-side system have been established above. The resulting water-side condition metric (W) results from the intersection of the unit condition observations and metrics. Thus,

$$W = PU \cap B \cap D \cap R \cap T, \tag{11.23}$$

where the unit condition observations and metrics; PU, B, D, R and T are obtained from Equations (11.12), (11.16), (11.17), (11.18) and (11.22), respectively.

Considering the form of D given by Equation (11.17), then Equation (11.23) may be reduced without loss of information to

$$W = PU \cap B \cap R \cap T. \tag{11.24}$$

Table 11.6 shows the combinations of fuzzy sets that are implied in this expression for W.

The eight other possible combinations involve $0.2PU_5$, which are all eliminated in the logic union operation and therefore need not be considered further.

In Table 11.6, column 9 shows the minimum of the active membership values in each row. Column 10 shows the subscript of the set associated with the minimum in each row. It is assumed in this case that a preponderance of subscript 4 sets in a row results in a W_4 set, whilst a preponderance of subscript 5 sets results in a W_5 set. If there is an equal balance of subscripts in rows 4, 6 and 7, it is only in the case of row 4 that the intersection result is affected. In the present case, if the subscript is interpreted as 4, then the final result is

$$W = 0.5028W_4 \cup 0.4W_5. \tag{11.25}$$

But if the subscript is interpreted as 5 the final result becomes

$$W = 0.5028 W_4 \cup 0.436 W_5. \tag{11.26}$$

The defuzzified value of Equation (11.25) is def $= 0.8608$. The defuzzified value of Equation (11.26) differs by only 0.62% from this value. Expression (11.25) [or (11.26)] represents the result of the water side condition audit for the heat exchanger system illustrated in Figure 11.1. Periodic condition auditing provides a means of tracking the system gross condition in time.

11.5. Value at Risk (VaR)

At management level, the system condition is inevitably related to a conceived VaR, which can have different interpretations depending upon the management objectives. Consider a prosaic example; the VaR of a used vehicle. The all-round satisfaction, as conceived by the prospective purchaser and which includes the vehicle's notional condition, would be compared with the value offered by the other alternatives within a selected price range. In all cases the value would be directly related to its notional condition. But the prospective insurance company for the same vehicle would consider the vehicle write-off value, the prospective purchaser's risk profile and other general accident associated liabilities, as the total VaR. Similarly, the VaR of practically every other system from nuclear power plant to health care, is open to various interpretations. Further reading on VaR may be found in in Glasserman [7].

11.6. Conclusion

Condition monitoring is widely practised and provides valuable data about a system's state and if it is conducted on a well organised basis, also about its historic pattern. The other important aspect is that of monitoring the corresponding system operating and environment conditions. This is not so frequently encountered, but is likely to influence the current condition of a system.

It is shown in this work how some of the principles of fuzzy logic may be applied to system condition monitoring observations to provide a system condition audit which results in an overall system condition metric.

The methodology provides additional system knowledge, which is useful at the strategic decision making level. The overall system condition metric would naturally be linked to a VaR at management level, reflecting management objectives. An improvement in knowledge management and presentation results from the adoption of the methodology advocated in this work.

System condition monitoring rarely includes the human operator skills, though it is acknowledged that this often has a profound influence on the condition of a system, whether it be of a physical or an intangible service type. There may be circumstances where the external boundary should be defined to include the human operator element.

The state of deterioration of the system is naturally expressed as the complement of its condition metric.

11.7. Notation

B	by-pass fuzzy condition metric
c_f	ideal filter constant
c'_f	observed filter constant
c_h	ideal pump characteristic
c'_h	observed pump characteristic
d	characteristic pump size
D	fuzzy control valve condition observation
E	fuzzy valve ease of operation
f	filter condition metric
F	fuzzy filter condition observation
g	acceleration due to gravity
G	fuzzy gland condition observation
h	pump head
i	pump condition observation, defined by Equation (11.5)
k_1	Stanton number coefficient
k_2	Reynolds number coefficient
L	fuzzy valve gland leakage observation
n	impeller rotational speed
p	pressure
P	fuzzy pump condition observation
PG	fuzzy pump and gland condition metric
PU	fuzzy pump unit condition metric
q	liquid volumetric flow rate
r_e	Reynolds number
R	fuzzy non-return valve condition observation
s_t	ideal Stanton number
s'_t	observed Stanton number
t	heat exchanger condition observation
W	liquid side overall fuzzy condition metric
x	general universe of discourse
X_i	ith general fuzzy set
Δ	an increment of ...
η	head load
$\mu(x)$	membership function
θ	temperature

∪   logic union
∩   logic intersection

References

1.   Geiker, M., "Durability Design of Concrete Structures", in *Minimising Total Life Costs – Considerations and Examples*, Proc. Intnl. Conf. on Concrete Durability and Repair Technology, September 1999, Univ. Dundee, Scotland, R.K. Dhio and M.J. McCarthy (eds), Th. Telford, UK, 1999, pp. 319–333.

2.   Harris, J., "Fuzzy Logic Methods in Fatigue and Creep", *Jour. Strain Anal.*, Vol. 36, No. 4, 2001, pp. 411–420.

3.   Harris, J., "Piecewise Linear Reliability Data Analysis with Fuzzy Sets", *Proc. Instn. Mech. Engrs.*, 2001, Vol. 215, Part C, 2001, pp. 1075–1082.

4.   Varona, J.M., Gutierrez - Solana, F., Gonzalez, J.J. and Gorrochategui, I., "Fatigue of Old Metallic Railroad Bridges", in *Intnl. Symp. on Fat. Design*, ESIS Publication 16, J. Sohn et al. (eds), Mechanical Engineering Publications, London, 1993, pp. 121–135.

5.   Wang, L.R. and Rowlands, H., "The Evaluation and Control of SPC in Fuzzy Logic and Neural Networks", in *2nd Workshop on Euro. Sci. and Indust. Collaboration (WESIC'99)*, Newport, Gwent, UK, pp. 391–398.

6.   Harris, J., "Reframing Standards Using Fuzzy Sets for Improved Quality Control", *Proc. Instn. Mech. Engrs.*, Vol. 16, 2001, pp. 91–98.

7.   Glasserman, P., "The Quest for Precision through Value-at-Risk", in *Mastering Risk, Financial Times*, 16 May 2000, Part 4, pp. 6–7.

8.   Harris, J., *An Introduction to Fuzzy Logic Applications*, Kluwer, Dordrecht, 2000.

9.   Davidson J. (ed.), *The Reliability of Mechanical Systems*, Mechanical Engineering Publications, London, 1994.

10. Andrews, J.D. and Moss, T.R., *Reliability and Risk Assessment*, Longman Scientific and Technical (now Pearson Education Limited), London, 1993.

Chapter 12
Reframing Standards Using Fuzzy Sets for Improved Quality Control *

Keywords. British Standard, Quality, Ultrasonics, Steel Castings, Fuzzy Sets.

Summary. British Standard BS 6208:1990 treats the ultrasonic inspection of steel castings and the consequent assignment of quality levels. This Standard is considered as a case study for reframing such specifications in terms of fuzzy sets. The advantages of using fuzzy sets are the elimination of sharp boundaries between quality levels and between cross-section zones with their attendant borderline cases, thus producing smooth natural changes with improved quality control. The Standard does not specify a method for estimating a consolidated quality level rating from the individual quality levels obtained from the several criteria in the Standard. Such a method, consequent on the fuzzy sets treatment, is proposed here.

12.1. Introduction

One of the hall marks of current activity in all sectors of society, particularly in the manufacturing and service sectors, is the drive towards continuous improvement of quality. The concept of quality embraces a wide variety of product (goods and services) properties, and it effectively defines the user satisfaction levels. Quality becomes an inherent feature from the initial concept stage, through all the subsequent stages of production or formulation to delivery and use, including servicing and maintenance support. It is therefore clear that every quality improving change in operations must be considered as likely value-enhancing.

The various modern paradigms of production are often applied as essentially quality improving tools, amongst them fuzzy logic is becoming increasingly familiar. There are many examples of this; Chen [1] applies fuzzy logic at the engineering design stage, Hinduja and Xu [2] consider tolerances and surface finishes associated with roughing, semi-finishing and finish machining operations. In the textile field Lin, Lin and Tsai [3] use fuzzy logic for fabric defects diagnosis

---

* This material has been reproduced from the Proceedings of the Institution of Mechanical Engineers, Part B, *Journal of Engineering Manufacture*, 2001, Volume 215 (B3), pp. 315–322, "Reforming standards using fuzzy sets for improved quality control", by J. Harris, with permission of the Council of the Institution of Mechanical Engineers.

and another different application was published by Harris [4] on raw milk quality assessment.

Published national and international standards provide a quality reference for both producers and users, and embody consensus opinion about best practice, which is a compromise of the significant factors of production and use of a product. Standards present product specifications and codes of practice as well as guidelines and glossaries to aid acceptable product quality, which includes reliability, maintainability and many other factors. In the foodstuffs industry nutritional value and hygiene are also important. It is obvious that standards can exhert a decisive influence on product quality, to list just a few.

Standards are often cast in terms of numerical ranges of parameters with hard boundaries, BS 6208:1990 which considers the inspection of ferritic steel castings by ultrasonics, is typical in this respect. It is taken as an example in the reframing of Standard specifications in terms of fuzzy sets. The interpretation of physical data affects the decisions on quality as much as the actual data and can lead to the rejection of products that might be acceptable and vice versa. Also the problem of borderline cases is removed and therefore also the human element in the judgement of such cases, which according to Fletcher [5] is the weakest link in a quality assessment system. Recasting Standards rules in terms of fuzzy sets provides a more natural system without the artificial sharp boundaries in the present form of the Standard.

12.2. Method of Ultrasonic Testing of Ferritic Steel Castings Including Quality Levels. BS 6208:1990

An important feature of BS 6208:1990 is the concept of zoning the casting wall section into a sandwich of two equal outer zones (OZ) separated by a mid zone (MZ) with sharp boundaries, as illustrated in Figures 12.1(a) and 12.1(b) for a varying wall thickness. The zones are defined in the Standard as follows:

(i) Wall thicknesses up to and including 30 mm. The full wall thickness is designated "the outer zone".
(ii) Wall thicknesses greater than 30 mm. The wall thickness is divided into layers designated outer and mid zones. The outer zones are defined as the lesser of

    (a) one third of the casting section thickness, or
    (b) 30 mm depth.

It will be noted in Figures 12.1(a) and 12.1(b) that there is an anomaly in the zone depth for a section thickness of 30 mm. Clarification has been sought on this matter, Harris [6].

In the published Standard, the defects detected by ultrasonic means are divided into planar defects and non-planar defects, the latter are evaluated in the zones

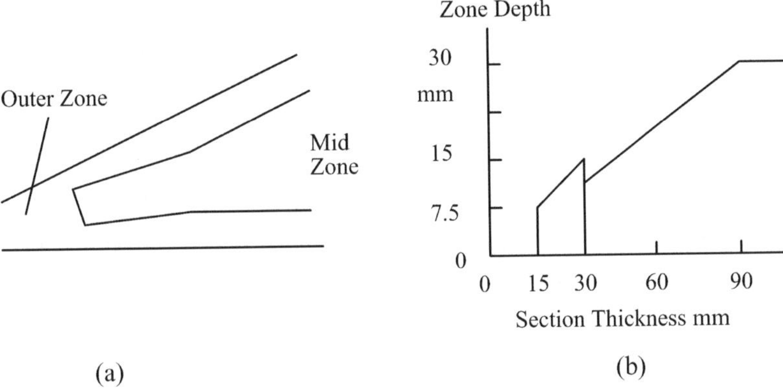

Figure 12.1. Zonal definitions (BS 6208:1990). (a) Illustration of zones for varying section thickness. (b) Outer zone depth v. section thickness (not to scale).

Table 12.1. Planar defect limits.

Criterion	Quality Level (QL)			
	1	2	3	4
Maximum indicated through-wall length of a single defect mm	0	5	8	11
Maximum indicated area of a single defect mm^2	0	75	200	360
Maximum total area of defects mm^2	0	150	400	700

described above. The four quality levels (QLs) in both cases are defined by the size and area of single defects and by their maximum total area. The QL criteria are summarised below in Tables 12.1 and 12.2 for ease of reference. The QL criteria for planar and non-planar defects imply that a casting may be assigned different quality levels according to the several criteria. This is discussed in more detail elsewhere, but in this work a method of finding a composite QL is proposed.

The test data for comparison with Tables 12.1 and 12.2 is not measured directly but is inferred from electronic recordings of sonic echoes. The methodology is indicated below.

12.3. Ultrasonic Defect Detection and Identification

In the ultra sonic method of inspection, sonic waves in the frequency range of 100 kHz to 20 MHz are generated by piezoelectric crystal probe are applied to the surface of the specimen, Fletcher [5]. The probe is moved across the surface of the component to provide a cross-section search (B-scan). This does not provide a direct representation of the cross-section and any present discontinuities, as in the case of an X-ray, but traces of wave intersections which include reflections,

Table 12.2. Non-planar defect limits.

Criterion	Quality Level (QL)			
	1	2	3	4
Outer Zone				
Maximum indicated single defect size (% of zonal thickness)	20	20	20	20
Maximum indicated defect area mm^2	250	1000	2000	4000
Maximum total defect t area mm^2	5000	10000	20000	40000
Mid Zone				
Maximum indicated single defect size (% of section thickness)	10	10	15	15
Maximum total defect area mm^2	12500	20000	31000	50000

refraction, diffraction, interference and modal conversion, which are analysed to form the basis of defect diagnosis.

The ultrasonic radiation within a specimen is uncollimated and therefore a point defect provides an arc trace which is normal to the incident wave beam direction at that point. The result of moving the probe across the specimen surface is to produce a set of intersecting arcs at the location of a point defect. A crack produces intersections with an approximate plateau, whilst a crack tip, a pore or an inclusion produce a more sharply defined location. The interpretation of the ultrasonic echoes requires significant diagnostic skill and to reduce variability of interpretation the task may be automated. An automated system is based upon an expert system with integrated artificial intelligence which incorporates a knowledge based system able to discount effects due to probe reverberations, back wall echoes and other false effects. The system knowledge base can also normally discriminate between symmetrical defects such as gas bubbles and asymmetrical flaws such as cracks. The latter are much more dependent upon the relative probe position. The knowledge base system identifies and integrates defect features detected by the ultrasonic waves and from the knowledge base forms a conclusion about its likely position, size, shape and nature (e.g., smooth or rough crack, inclusion or porosity), Halmshaw [7]. Groups of small inclusions may appear to be cracks and BS 6208:1990 specifies that defects less than 25 mm apart are infact to be considered as a single larger defect. This is to be accounted for in the QL ratings based upon defect size and maximum total defect area.

Some knowledge based systems include a learning function, which may be either supervised or unsupervised. This assists, but does not eliminate uncertainty.

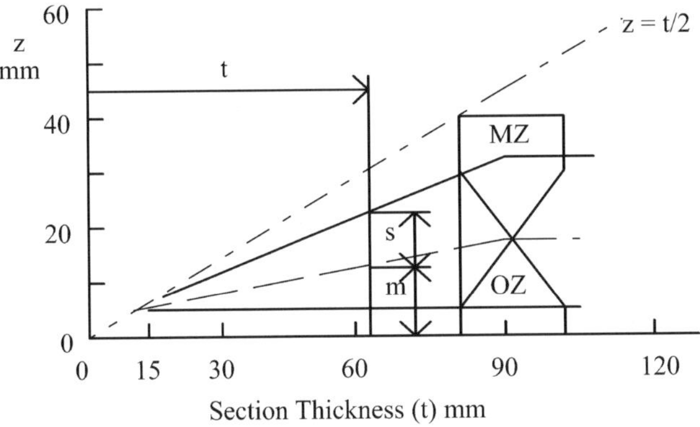

Figure 12.2. Fuzzy zoning of the casting section.

12.4. Reframing BS6208:1990

12.4.1. FUZZY DATA

It will be evident from the above discussion that ultrasonic detection, identification and classification of casting defects is not an exact science, but retains an element of uncertainty which is not eliminated by automation. This is not reflected at present in the reporting of defect parameters, which could be more realistically be reported in terms of fuzzy sets. The current edition of BS 6208:1990 is only suited to data in the form of crisp values, not fuzzy sets. In the following work attention is directed towards reframing BS 6208:1990 in terms that will accept either fuzzy or crisp input data. The method of treating fuzzy input data is given by Ross [8] and Harris [9] and is not persued here.

12.4.2. ZONE PARTITIONING

In BS 6208:1990 zoning of the casting cross-section is an important factor for non-planar defects in wall thicknesses above 15 mm, whilst for wall thicknesses less than 15 mm there is no zoning. Thus for a change in wall thickness from slightly less than 15 mm to slightly more than this value there is a step change in the treatment of non-planar defects. This is a prominent example of the problems caused by hard classification. The dichotomy is removed by defining two overlapping fuzzy sets, the extent of the overlap can be conveniently made proportional to the wall thickness between wall thickness limits of 15 mm and 90 mm in conformity with the limits given in the Standard. This is illustrated in Figure 12.2.

At a wall section thickness of 90 mm the overlap of OZ and MZ becomes $23.5 \pm$ 16 mm and thereafter is independent of the section thickness for larger values of t. This proposed pattern is considered to incorporate the essential features of the zoning pattern given in BS 6208:1990, but avoids the problems of artificial sharp boundaries. For a typical value of the section thickness the zone composition would

Table 12.3. Fuzzy zonal membership values.

t (mm)	$\mu()$	z (mm)								
		7.5	10	12.5	15	20	25	30	35	40
15	OZ	1								
	MZ	0								
30	OZ	1	0.609	0.219	0					
	MZ	0	0.391	0.781	1					
45	OZ	1	0.805	0.609	0.414	0.024	0			
	MZ	0	0.195	0.391	0.586	0.976	1			
60	OZ	1	0.870	0.740	0.609	0.350	0.090	0		
	MZ	0	0.130	0.260	0.391	0.650	0.910	1		
75	OZ	1	0.902	0.805	0.707	0.512	0.316	0.121	0	
	MZ	0	0.098	0.195	0.293	0.488	0.684	0.879	1	
90	OZ	1	0.922	0.844	0.766	0.609	0.453	0.297	0.141	0
	MZ	0	0.078	0.156	0.234	0.391	0.547	0.703	0.859	1

comprise an outer zone with a membership value ranging from unity at the surface to a depth of 7.5 mm, then decreasing uniformly to zero at a depth $z = (m + s)$. In the same range of t values the mid zone would increase from zero to unity. For all sections in the range of t between 15 and 90 mm, m is the cross-over point for the OZ and MZ zones and $2s$ is their overlap. Different interpretations of the Standard are possible, but the one proposed here in the range of section thicknesses between 15 and 90 mm is

$$s = 16(t - 15)/75 \text{ mm} \tag{12.1}$$

and

$$m = (s + 7.5) \text{ mm}. \tag{12.2}$$

At any section in the above range the membership values (μ) for the OZ and MZ are given by

$$\mu(OZ) = (m + s - z)/(m + s - 7.5) = (m + s - z)/25, \tag{12.3}$$

$$\mu(MZ) = 1 - m(OZ). \tag{12.4}$$

Values are tabulated in Table 12.3 for reference.

Fuzzy membership values (μ) for OZ and MZ are tabulated for various depths (z) and for a range of casting section thicknesses (t) in Table 12.3, whilst corresponding values of s and m are tabulated in Table 12.4.

Table 12.4. Values of m and s.

t (mm) m (mm)	s (mm)	$(m + s)$ mm
15	7.5	0	7.5
30	10.7	3.2	13.9
45	13.9	6.4	20.3
60	17.1	9.6	26.7
75	20.3	12.8	33.1
90	23.5	16	39.5

Table 12.5. QL membership values for planar defect length.

	Defect Length mm							
QL	0	2	4	5	6	8	10	11
1	1	0.5	0					
2	0	0.5	1	0.5	0			
3			0	0.5	1	0.5	0	
4					0	0.5	1	1

12.4.3. QUALITY LEVEL ASSESSMENT FOR PLANAR DEFECTS

Quality levels in the published BS 6208:1990 are defined in terms of crisp sets with the effect that a slight change in a criterion (e.g. maximum through wall length of a single defect) around a boundary value can make a step change in the quality level. This makes calibration of test equipment and uniformity of interpretation, whether human or automatic, a critical issue. In view of the previous discussion this may make unreal demands, particularly if production is spread across one or more countries. Casting the QLs and the criteria as fuzzy sets reduces their criticality without impeaching their standard.

Considering the case of planar defects, and framing the defect length, defect area and total defect area as fuzzy sets requires a choice of partitioning on the universes of discourse. Here the cross-over points of the sets are chosen to reflect the quality limits in the Standard. Figures 12.3(a), 12.3(b) and 12.3(c) illustrate the resulting partitioning,

Membership values of the fuzzy sets illustrated in Figure 12.3 are shown in Tables 12.5, 12.6 and 12.7. (It is understood that blank entries are zeros.)

EXAMPLE. As an example of the application of the fuzzy formulation of planar defects, consider a defect with a measured length of 4.3 mm and an area of 23 mm^2. By interpolation of Table 12.5, the membership values are found as

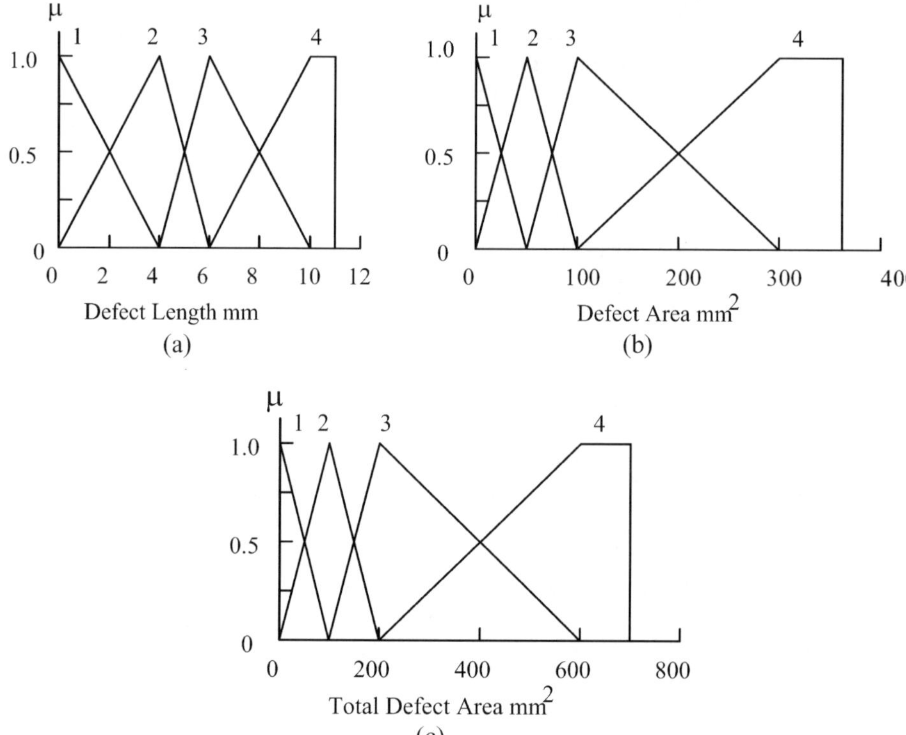

Figure 12.3. Fuzzified planar defect criteria. (a) Defect length. (b) Defect area. (c) Total defect area.

Table 12.6. QL membership values for planar defect area.

	Defect Area mm^2							
QL	0	25	50	75	100	200	300	350
1	1	0.5	0					
2	0	0.5	1	0.5	0			
3			0	0.5	1	0.5	0	
4					0	0.5	1	1

follows:

For the defect size of 4.3 mm:

$$\mu(\text{QL2}) = 1 - 0.5 * 0.3 = 0.85$$

and

$$\mu(\text{QL3}) = 1 - \mu(\text{QL2}) = 0.15.$$

Table 12.7. QL membership values for planar total defect area.

	Total Defect Area mm^2							
QL	**0**	**50**	**100**	**150**	**200**	**400**	**600**	**700**
1	1	0.5	0					
2	0	0.5	1	0.5	0			
3			0	0.5	1	0.5	0	
4					0	0.5	1	1

Therefore the resultant QL is (0.85QL2, 0.15QL3). This may be defuzzified to give a single value which is fractional. It may be noted that the QL is biased towards QL2 with a smaller tendency towards QL3. Using the centroid method, the result is

$$QL = (0.85 * 2 + 0.15 * 3)/(0.85 + 0.15) = 2.15.$$

For the defect area of 23 mm^2:

$$\mu(QL1) = 1 - 0.5 * 23/25 = 0.54$$

and

$$\mu(QL2) = 1 - \mu(QL1) = 0.46.$$

The resultant QL in this case is (0.54QL1, 0.46QL2), which may also be defuzzified as before, giving

$$QL = 1.46.$$

The defuzzified QL forms found above are useful for forming a composite QL as will be shown later.

12.4.4. QUALITY LEVEL ASSESSMENT FOR NON-PLANAR DEFECTS

Unlike the case of planar defects, non-planar defect QLs are influenced by their depth within the casting, the partitioning of the cross-section discussed in Section 4.2 is therefore important. It will be noted that the maximum defect size criteria for non-planar defects in Table 12.2 are given as a percentage of the zonal thickness for the outer zone and as a percentage of the section thickness for the mid zone rather than directly as for planar defects. For the outer zone the percentage is the same for all QLs (20% of zonal thickness) and hence fuzzification is not relevant. For the mid zone QL1 = QL2 (10% of section thickness), also QL3 = QL4 (15% of section thickness), so there is one step change which could be fuzzified into two sets, but it would not make a significant change to the outcome. Hence only the

Table 12.8. Minimum and maximum non-planar defect sizes.

			Defect Size mm (D)			
t mm	**($m - s$) mm**	**($m + s$) mm**	**QL1**		**QL3**	
			Min	**Max**	**Min**	**Max**
15	7.5	7.5	1.5	1.5	1.5	1.5
30	7.5	13.9	2.14	3.00	2.14	3.75
45	7.5	20.3	2.78	4.50	2.78	6.00
60	7.5	26.7	3.42	6.00	3.42	8.25
90	7.5	39.5	4.70	9.00	4.70	12.75

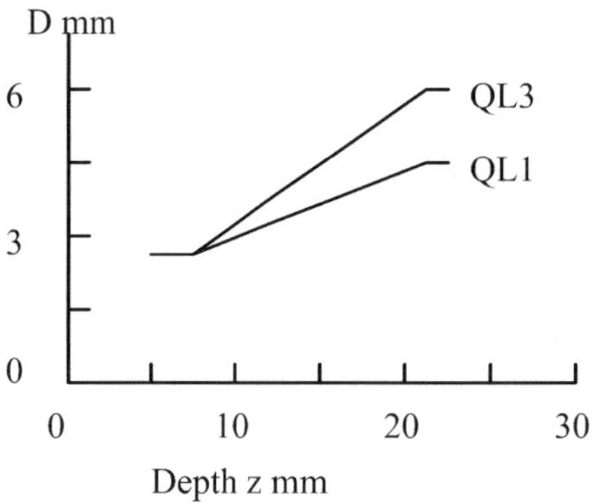

Figure 12.4. Non-planar defect size against casting depth for a casting wall section of 45 mm.

effect of fuzzy zoning will be considered. The defect size criteria may therefore be expressed as follows:

$$\text{For QL1 (QL2)} \quad D = 0.2m\mu(\text{OZ}) + 0.1t\mu(\text{MZ}), \tag{12.5}$$

$$\text{For QL3 (QL4)} \quad D = 0.2m\mu(\text{OZ}) + 0.15(t - 5)\mu(\text{MZ}), \tag{12.6}$$

where the membership values $\mu(\text{OZ})$ and $\mu(\text{MZ})$ are obtained from Figure 12.2 or are interpolated from Table 12.3. The -5 factor is included in the right-hand side of Equation (12.6) to avoid the anomaly present in the profiles shown in Figure 12.1.

Since in Equations (12.5) and (12.6) the membership values on the right-hand side are linear in z and m is a constant for a given section, then a table of minimum and maximum values of D for a range of values of t contains sufficient information for $D(z)$ to be evaluated. Sample values are given in Table 12.8 and are illustrated in Figure 12.4.

EXAMPLE. Consider a non-planar defect of 4.32 mm at a depth of 17 mm in a casting section thickness of 45 mm. The appropriate QL may be evaluated by finding the membership values of MZ and OZ, then calculating the D values using Equation (12.5) for QL1 (or QL2) and Equation (12.6) for QL3 (or QL4). Finally the measured D may be compared with the calculated values to ascertain the membership of the adjacent quality levels.

Thus, the membership values of MZ and OZ are first calculated by noting from Table 12.8 (for example) that $(m + s)$ is 20.3 mm also that $(m - s)$ is 7.5 mm. Hence at a depth of 17 mm,

$$\mu(\text{MZ}) = (17 - 7.5)/(20.3 - 7.5) = 0.742$$

and

$$\mu(\text{OZ}) = 1 - 0.742 = 0.258.$$

From Equations (12.5) and (12.6) (noting that $m = 13.9$ mm)

$$D(\text{QL1}) = 0.2 * 13.9 * 0.258 + 0.1 * 45 * 0.742 = 4.056 \text{ mm}$$

and

$$D(\text{QL3}) = 0.2 * 13.9 * 0.258 + 0.15 * (45 - 5) * 0.742 = 5.169 \text{ mm}.$$

Now comparing the measured value D' of 4.32 mm,

$$\mu(\text{QL1}) = (5.169 - 4.32)/(5.169 - 4.056) = 0.763$$

and

$$\mu(\text{QL3}) = 1 - 0.763 = 0.237.$$

Hence the QL of the given defect is (0.763QL2, 0.237QL3). (Note, QL1 = QL2.) It is possible to defuzzify the QL using the centroid method giving

$$QL' = 0.763 * 2 + 0.237 * 3 = 2.237.$$

This value will be used later.

12.4.5. QUALITY LEVEL ASSESSMENT FOR NON-PLANAR DEFECT AREAS

BS 6208:1990 specifies QLs for the maximum indicated defect areas in the outer zone, but not in the mid zone. The fuzzified chart is shown in Figure 12.5. If the QLs for the mid zone were also specified by agreement between supplier and customer, then the principles outlined below could be applied to obtain the QL distribution across the casting section.

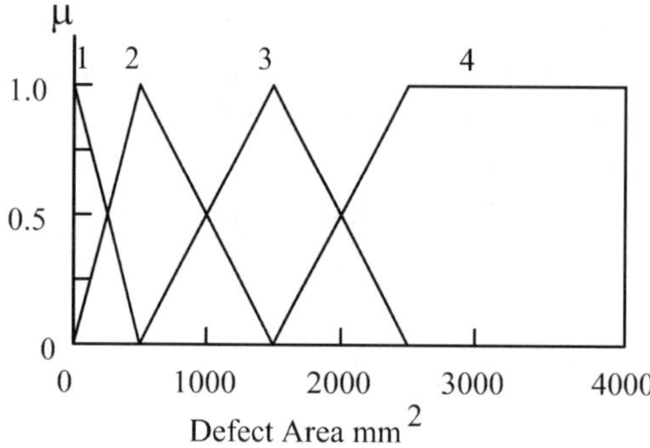

Figure 12.5. Fuzzified non-planar defect area criteria.

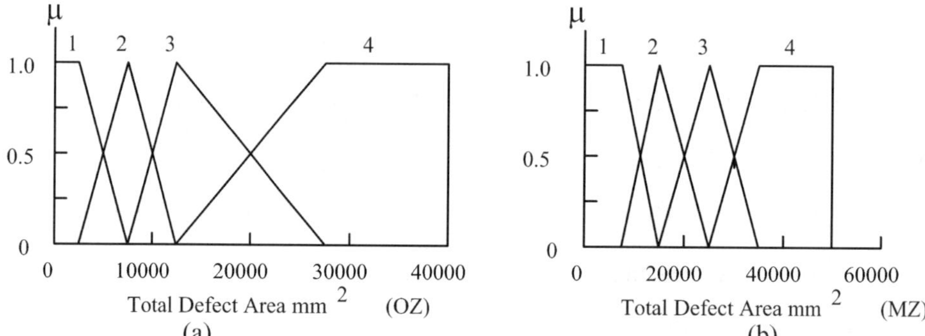

Figure 12.6. Fuzzified non-planar total defect area. (a) Outer zone. (b) Mid zone.

12.4.6. QUALITY LEVEL ASSESSMENT FOR NON-PLANAR TOTAL DEFECT AREA

Following the same pattern as given above, the maximum total defect area rules for QLs in Table 12.2 can also be reframed as fuzzy sets. Figures 12.5(a) and 12.5(b) illustrate the partitioning of the total defect area that is compatible with the Standard specification. For a given total defect area, the fuzzy set diagrams in Figures 12.6(a) and 12.6(b) below will each be intersected twice yielding membership values μ_{oa} and μ_{ob} for Figure 12.6(a), and μ_{mc} and μ_{md} for Figure 12.6(b). Then weighting the OZ and MZ QLs with the zonal membership values $\mu(\text{OZ})$ and $\mu(\text{MZ})$ gives the following expression for the resultant QL:

$$QL = \mu(\text{OZ})(\mu_{oa}QL_a, \mu_{ob}QL_b) + \mu(\text{MZ})(\mu_{mc}QL_c, \mu_{md}QL_d), \qquad (12.7)$$

where QL_a, QL_b, QL_c, and QL_d are the corresponding quality levels.

Membership values for the fuzzy sets illustrated in Figure 12.6(a) and 12.6(b) are shown in Tables 12.9 and 12.10. (It is understood that blank entries are zeros.)

The application of Equation (12.7) is illustrated in the following example.

Table 12.9. QL membership values for OZ non-planar total defect areas.

QL	\multicolumn{9}{c}{Total Defect Area $* 10^{-4}$ mm^2}

QL	0	0.25	0.5	0.75	1.0	1.25	2.0	2.75	4.0
1	1	1	0.5	0					
2		0	0.5	1	0.5	0			
3			0	0.5	1	0.5	0		
4					0	0.5	1	1	

Table 12.10. QL membership values for MZ non-planar total defect areas.

QL	0	0.875	1.25	1.625	2.0	2.375	3.100	3.825	5.0
1	1	1	0.5	0					
2		0	0.5	1	0.5	0			
3			0	0.5	1	0.5	0		
4					0	0.5	1	1	

EXAMPLE. As a representative calculation of how the QL fuzzy sets are used in conjunction with the zoning shown in Figure 12.2, consider a casting section thickness of 45 mm with a total defect area of $2.148 * 10^{-4}$ mm^2 at a depth of 15 mm from the surface.

By interpolation of Table 12.9 the OZ QL3 membership value $= 0.5 - 0.5 * 0.148/0.75 = 0.4$. Hence the OZ QL4 membership value is $1 - 0.4 = 0.6$.

By interpolation of Table 12.10 the MZ QL2 membership value $= 0.5 - 0.5 * 0.148/0.375 = 0.303$. Hence the MZ QL3 membership value is $1 - 0.303 = 0.697$.

The outer zone QL is therefore expressed by (0.4QL3, 0.6QL4). The inner zone QL is similarly expressed by (0.303QL2, 0.697QL3). From Table 12.3, at a depth of 15 mm in a 45 mm thick casting section the membership values are $\mu(\text{OZ}) = 0.414$ and $\mu(\text{MZ}) = 0.586$.

Hence the overall QL is given by

$$QL = 0.586(0.303QL2) + (0.414 * 0.4 + 0.586 \times 0.697)QL3 + 0.414(0.6QL4)$$

$$= 0.178QL2 + 0.574QL3 + 0.248QL4.$$

This quality level may be simplified by defuzzifying using the centroid method giving

$$QL' = (0.178 * 2 + 0.574 * 3 + 0.248 * 4)/(0.178 + 0.574 + 0.248)$$

$$= 3.070.$$

Thus QL is slightly greater than QL'3.

12.5. Compound Quality Level

In the above work a number of quality levels have been found for planar and non-planar defects. The question then arises about the overall quality level of the casting. BS 6208:1990 does not give any guidance on this aspect. It is therefore proposed that an overall value may be found for n QLs by taking the nth root of their product. Thus for the examples used above the following QLs have been found:

Planar defects:	Defect size	2.15
	Defect area	1.46
Non-planar defects:	Defect size	2.237
	Total defect area	2.898

The consolidated QL is therefore,

$$QL'' = (2.15 * 1.46 * 2.237 * 3.070)^{1/4}$$

$$= 2.150.$$

This lies between QL2 and QL3, and is biased towards QL2 with a small tendency towards QL3.

12.6. Conclusion

It is shown in the above work how the specification of quality of steel castings based upon ultrasonic testing may be formulated in terms of fuzzy sets. The British Standard BS 6208:1990 is used as the reference for this and it represents a case study for reframing standards. The fuzzy set type of formulation avoids step changes and hard boundaries between classes and is therefore a more natural representation of parameter ranges. The difficulties associated with borderline cases is avoided and therefore also the associated discrepancies which may arise from different product sources. Quality control is improved in this way and also by having a more refined smooth gradation of zones within the material rather than just two exclusive zones as is the present case.

The methods outlined in this work can be incorporated in an automated system, whilst for non-automated systems look-up tables of values could be provided which make no additional demands on the skills of the operator and would infact require no decision making for border line cases. There is more work involved with the above treatment compared with the current Standard, but this would be hidden from the operator if it was programmed for machine operation, then the output would simply be quality levels.

In BS 6208:1990 there is an anomaly at a section thickness of 30 mm, the above treatment eliminates this.

References

1. Chen, Y.H., "Fuzzy Ratings in Mechanical Engineering Design – Applications to Bearing Solutions", *Proc. Inst. Mech. Engrs.*, UK, Vol. 210 (B1), 1996, pp. 49–53.
2. Hinduja, S. and Xu, X., "Determination of Finishing Features in $2\frac{1}{2}$-D Components", *Proc. Inst. Mech. Engrs.*, UK, Vol. 211 (B2), 1997, pp. 125–142.
3. Lin, J-J., Lin, C-H. and Tsai, I.S., "Applying Expert System and Fuzzy Logic to an Intelligent Diagnostic System for Fabric Inspection", *Textile Research Jour.*, Vol. 65(12), 1995, pp. 697–709.
4. Harris, J., "Raw Milk Grading Using Fuzzy Logic", *Intnl. Jour. of Dairy Techn.*, Vol. 51(2), 1998, pp. 54–56.
5. Fletcher, M.J., "Non-Destructive Testing", in *Kempes Engrs. Year-Book*, 1999, D2/1089-1106, ISBN 0-86382-408-0.
6. Harris, J., Private Communications to the British Standard Institution dated 27/11/99 and 31/11/99.
7. Halmshaw, R., *Non-Destructive Testing*, Edward Arnold, 1987, ISBN 0-713-13634-0.
8. Ross, J.R., *Fuzzy Logic with Engineering Applications*, McGraw-Hill Inc., 1995, ISBN 0-07-053917-0.
9. Harris, J., *An Introduction to Fuzzy Logic Applications*, Kluwer Academic Publishers, 2000.

Chapter 13
On the Correlation of Statistical and Automatic Process Control[*]

Keywords. Control chart, Fuzzy logic, FTA, Gauss, Zone rules, Random variations, Assignable variations.

Abstract. A process control strategy is proposed based upon the twin themes of statistical and automatic process control. The main categories of product fault are identified and related to the capabilities of statistical and automatic control. Statistical control is supported by process fault information from a process specific fault tree analysis, which provides the basis for a corrective intervention protocol. Application is discussed in terms of fuzzy automatic control, which offers a greater generality than conventional automatic control modelling. Prior publications which fuzzifies statistical control zones are arguably incomplete in the application of logic propositions and also in the identification of process faults. The present work proposes a general strategy, which may be adapted to specific processes. Both control by variables and control by attributes may be included within this treatment.

13.1. Introduction

Statistical process control (SPC) and automatic control have had historically separate paths. The former originated in the 20th century in light engineering mass production quality control and the latter in the 19th century in servo-mechanisms, an example of which is the well-known Watt govenor. In automatic control the two main concerns are system tracking and system regulation, and a substantial corpus of theory has been published encompassing both linear and non-linear systems, the most general being for linear systems. Tracking and regulation studies differ mainly in emphasis tather than fundamentals. In the present context, process regulation is principal interest.

The ideal of maintaining a product stream which is always within specification, be it comprising components, assemblies, bagged, packaged or containerised

---

* This material has been reproduced from the Proceedings of the Institution of Mechanical Engineers, Part B, *Journal of Engineering Manufacture*, 2003, Volume 217 (B1), pp. 99–109, "On the correlation of statistical and automatic process control", by J. Harris, with permission of the Council of the Institution of Mechanical Engineers.

solids, fluids or particulate matter from continuous or batch processes is generally not easy to achieve. The product stream is subject to a range of influences such as:

(i)   Input materials specification,
(ii) Process function and state variability,
(iii) Environmental conditions,
(iv) Operator human factors and variability,
(v)   Reliability and quality of support services such as maintenance.

SPC is a well-known and widely publicised tool which is employed in pursuance of product specification compliance through the regulation of the production process. It will be obvious from the above list that SPC is generally effected by external intervention, unlike automatic process control (APC) which is generally a closed-loop and on-line sub-system of the process. These characteristics are incorporated into the control strategy outlined in this work.

Control charts play a prominent role in SPC applications and although they originated in the light engineering sector, their use has since spread to other areas of production and processing, such as the chemical industry. The control chart is conventionally demarcated by upper and lower control limits (UCL and LCL respectively), which are often symmetrically placed about a central value line.

In the application of control charts, Davies [6] has emphasised that each case needs to be considered in detail and treated on its own merits to determine the appropriate form of control chart. This is also true in the wider context of SPC-APC applications. At the commissioning stage it must be established in the first place, whether or not the system is capable of providing an output of the requisite specification at the desired rate. If it is, then some of the control function may be exercised by APC. Those parts of the control function that cannot be managed under APC, because of the type or range of action required, must be managed under SPC and external intervention such as corrective maintenance or change of inputs, for example. APC may, or may not, share a common sampling method with the associated SPC.

In batch processing, steady state is not achieved and therefore successive samples are not drawn from the same universe and are not statistically uniform, though sample analysis may, of course, be compared with a specification. But Davies [6] has drawn attention to the questionable validity of statistical conclusions drawn under such circumstances.

For any control system of whatever type, the time-scale of the control (including the sampling system) response must be of a smaller order of magnitude than that of the production process for it to be effective. The process may need to be stopped for SPC intervention, but APC intervention is normally on-line although it is known that conventional linear APC can sometimes be applied to non-linear systems by tuning using the Ziegler–Nichols rules, see for example Thompson [7], a fuzzy logic representation is used in this work, providing fuzzy automatic process control

(FAPC). Such a representation may be applied even when no mathematical model of the system is known.

13.2. Statistical Process Control

13.2.1. BASIC CONCEPTS

If a system is under control it means that its output is statistically uniform and that the product is all from the same universe. This does not mean that the product mostly complies with its quality specification. For control by variables the quality will be in terms of a critical parameter target value (with a tolerance range), typically a dimension, strength, weight or electrical resistance, for example. The observed average of the output samples may not coincide with the target values, or even if it does, the scatter in the values may be too great. The level of process control must match the quality specification. If the former type of fault is noted, it may be corrected by process adjustment under APC or SPC, but if the latter type of fault occurs it may entail more fundamental action with off-line intervention, perhaps up to system redesign. An alternative is to amend the specification to match the level of control available.

In any application of SPC the hypothesis being tested is that any detected variation in the data is only due to stochastic processes. Two types of variation may be observed in sample data:

(i) Chance causes. These are due to unassigned events within the process or small sudden variations in environmental conditions, in inputs or in operator actions.
(ii) Assignable variations. These are due to accountable causes and usually produce identifiable patterns on a control chart. They are produced, for example, by process deterioration, mechanical faults, a change in source of raw material or process operator fatigue.

The effects of both the above types of variation are additive. In the unassignable chance causes category, part of this may be due to random effects which can be termed "noise". It is well-known that integration reduces the effect of signal noise and similarly, the effect of averaging sample values each of say, 5–10 production units, will assist in mitigating the effects of random or periodic variations. Clearly, the larger the sample size, the greater the reduction, but subject to control system and economic efficiency. SPC cannot be applied to chance causes. In the case of non-random variations in the system output, the identification and classification of causes is important. Frequently, in system design a fault tree analysis (FTA) is conducted which will provide vital information on the most likely causes of output variations. If this is not available, the first step is to create those FTAs that are relevant to the process. The FTAs will provide the most probable cause and effect relationships which will guide process control.

The SPC concepts outlined above are mainly concerned with the observation of output variables data, where individual measurements are made on specimen samples. The background theory is based upon a Gaussian (normal) probability distribution. Where the data is in the form of the number of defective items per output batch, the theory is based upon the binomial distribution, and where it is in the form of the number of defects per unit output (as in metal, plastic sheet or textile production), it is based upon the Poisson distribution. In the first case, this is termed analysis by variables, whilst in the latter two cases it is termed analysis by attributes. The SPC concepts are the same in all cases.

13.2.2. CONTROL CHARTS

The conventional use of control charts is as a tool to detect when observed variations of a critical given output parameter cease to arise solely from chance causes and deterministic patterns become apparent. If this occurs, the data is no longer statisically uniform and statisictal conclusions can no longer be reliably drawn. It is then assumed that within the sample data only chance causes of variation exist, but between sample groups assignable causes are present. Sample size and sampling frequency must be chosen with care to minimise internal variation, to represent all the process variations in the sample groups and to observe economic efficiency.

Normally, two types of chart are constructed: (i) a chart for sample averages and (ii) a chart for dispersion (either standard deviation or range). Typical control chart types are illustrated in Figures 13.1(a) and 13.11(b). Figure 13.1(a) is for sample average and shows outer control limits (OCLs) and inner control limits (ICLs). For a Gaussian (normal) probability, the OCLs are at $x_o = x_a \pm 3.06\sigma/\sqrt{n}$, whilst the ICLs are at $x_i = x_a \pm 1.96\sigma/\sqrt{n}$, where σ is the standard deviation of the universe, x_a is the central value and n is the sample size. With these limits, 5% of the samples will lie outside the ICLs, but only 0.2% will lie outside the OCLs. Sometimes a simpler type of chart is used with only two control limits called the upper and lower control limit (UCL and LCL respectively). These limits correspond closely with the position of the two OCLs.

Generally, the control limits may be expressed as

$$\text{Average:} \quad \text{OCL} \quad x_o = x_a \pm A(n)\sigma, \tag{13.1}$$

$$\text{ICL} \quad x_i = x_a \pm B(n)\sigma. \tag{13.2}$$

$$\text{Range:} \quad \text{OCL} \quad r_o = r_a \pm C(n)\sigma, \tag{13.3}$$

$$\text{ICL} \quad r_i = r_a \pm D(n)\sigma. \tag{13.4}$$

Tabulated values of $A(n)$, $B(n)$, $C(n)$ and $D(n)$ are available in Davies [6] and also for $A(n)$ and $C(n)$ for different probability distributions in Raz [11].

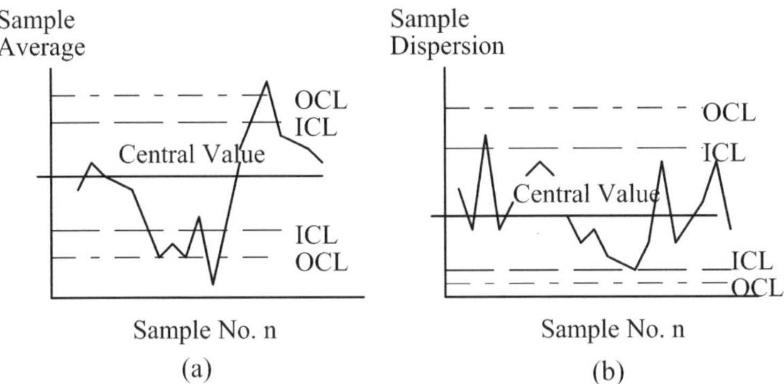

Figure 13.1. Illustration of typical control charts. (a) Sample averages. (b) Sample dispersions.

13.2.3. INTERPRETATION OF OBSERVATIONS

The simplest rule of SPC is that a sample value lying outside the control limits indicates a process out-of-control condition. If the causes of the observed fluctuations about the central values are stochastic, then sample values have a 50% probability of being on either side of the average. If a number of consecutive values fall on one side of the average, then there is a probability that there is some deterministic cause. For example, if seven consecutive values lie on one side of the average, the probability of this is 0.78%, Raz [10], which shows that this is an unlikely event. The probabilities of other patterns, based on a Gaussian distribution, can also be found. These are further indicators of process out-of-control conditions.

If the sample average values continues to be biased on one side of the control chart whilst the dispersion remains unchanged, then corrective action is indicated. The process itself is however likely to be controllable. Figure 13.1 shows control charts for a process that is out of control, but controllable. If, however, the dispersion is biased on the $+ve$ side, but the sample means remain symmetrical, then a more fundamental process fault has occurred which is unlikely to be controllable by APC, but requires off-line correction by repair or replacement maintenance, or perhaps process modification. If the dispersion is biased on the $-ve$ side of the average, it may be caused by a reduction in accuracy of the sampling and inspection procedure, which again may require off-line correction. It is clear from the above discussion that clusters of sample observations that are biased on one side or the other of the central value of the sample averages or dispersion indicates a process fault. This has been formalised by introducing zones within the OCLs at $\pm\sigma$, $\pm2\sigma$ and $\pm3\sigma$ on both the sample average and dispersion charts, as shown in Figure 13.2. For a Gaussian distribution, approximately 66.7% of the sample values lie in the 0–σ (C) zone, 95% lie in the 0–2σ ($C + B$) zones and 99.75% lie in the 0–3σ ($C + B + A$) zones.

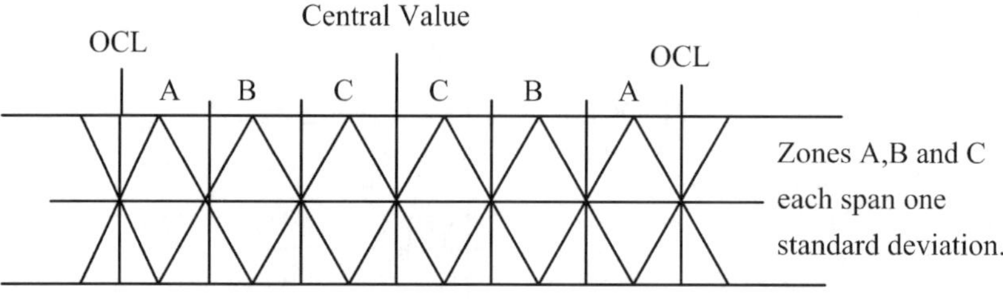

Figure 13.2. Zoning of the control limits. After Wang and Rowlands [5].

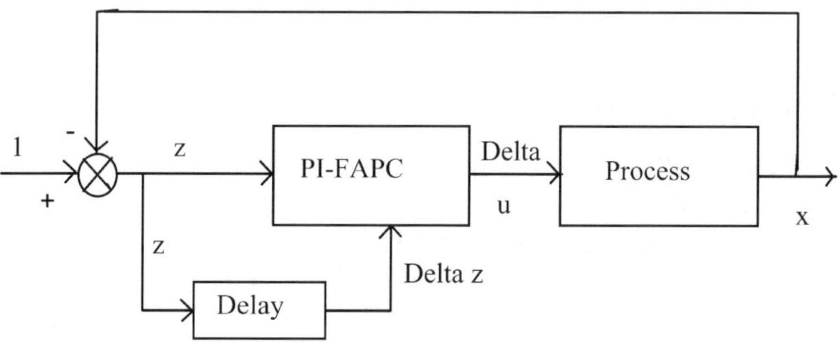

Figure 13.3. PI process control block diagram.

13.3. Fuzzy Automatic Process Control (FAPC)

For process regulation, proportional-integral control (PI) is commonly applied and it is the type which could be employed in the present context. It is emulated in fuzzy logic terms in the feed-back control described below. The block diagram in Figure 13.3 shows the principle of PI-FAPC.

Tracking can also be achieved by the incorporation of a certain degree of derivative control, but with the penalty of increasing instability. There is no general theory of non-linear process control, however, FAPC as described below, can be applied to any type of process, whether or not it is linear, or indeed has any known mathematical model. Sensitivity and stability are important concepts in APC and are discussed in standard control theory texts, see for example Thompson [7]. There is no available stability theory for SPC or FAPC, each case must be studied on its own merits.

Although both sample central values and dispersion are of interest in process control, it is noted above that out-of-control dispersion is most likely to be caused by a process fault requiring repair or replacement maintenance which is out of the province of FAPC. Therefore, in the following discussion only control of the average value is considered.

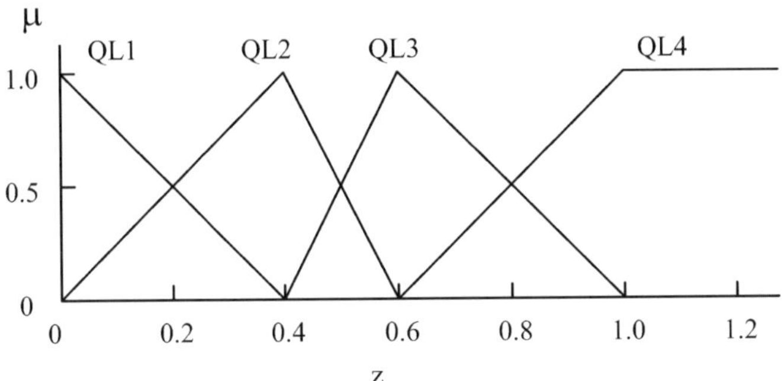

Figure 13.4. Partitioning of the z space.

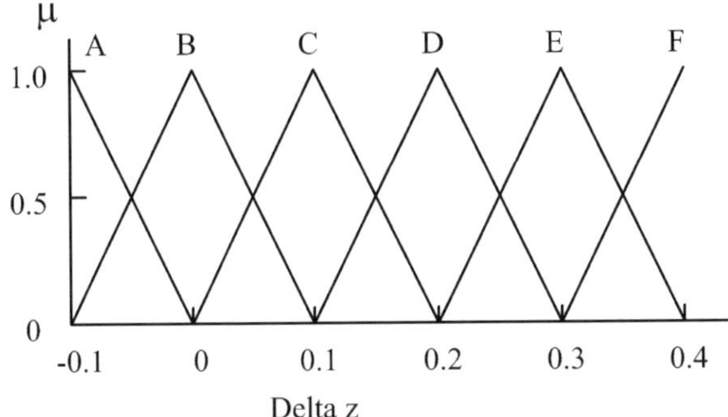

Figure 13.5. Partitioning of the δz space.

Harris [3] has discussed a typical second-order process model, which in finite difference form is given by

$$z_{n+1} = a\left(\sum \delta u_i + z_n - b z_{n-1}\right).\tag{13.5}$$

In the above equation, the δu_i are the incremental outputs of the PI fuzzy controller and z_n is given by

$$z_n = 1 - x_n,\tag{13.6}$$

where x_n is the scaled nth process output sample metric. This metric implies a specific out-of-control process state.

Representations of z, δz and δu spaces are shown in Figures 13.4, 13.5 and 13.6 respectively. The z and δz comprise the control inputs. The QL ranges reflect quality levels in BS 6208. Interpreting product faults as quality levels enables supplier penalties to be imposed.

A typical relational base for a PI type controller output is shown in Table 13.1.

The control process is governed by the operation of a fuzzy logic intersection form of proposition:

IF Z　AND δZ　THEN U　MIN　CONSEQUENCE

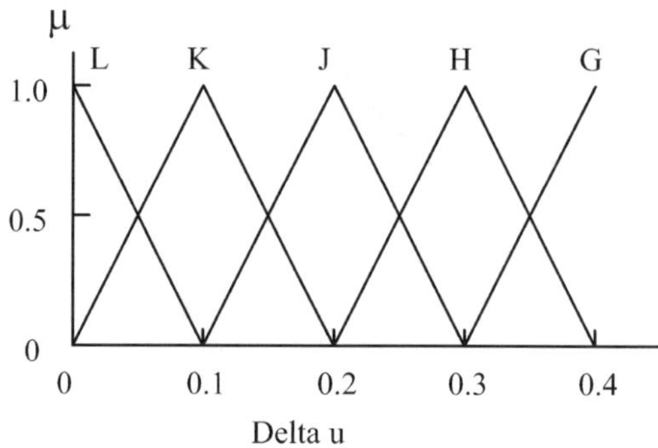

Figure 13.6. Partitioning of the δu space.

Table 13.1. Typical PI controller relational base.

| | δz | | | | | |
z	A	B	C	D	E	F
QL1	G	H	J	J	K	L
QL2	H	J	J	K	K	L
QL3	J	J	J	K	L	L
QL4	J	J	K	L	L	L

The operation is similar to the intersection operation in two-valued logic.

Since both Z and δZ yield two fuzzy sets each, there are four consequences, which by logical union, comprise the fuzzy conclusion, see for example Harris [3]. The fuzzy conclusion may be defuzzified to provide the numerical value of the control input signal to the production process.

13.4. Control Strategy

Consideration is now given to the correlation of SPC and FAPC for out-of-control production process states to provide an overall process control strategy. Each critical output parameter must be treated individually as described below.

There are four recognised types of control chart pattern that are assumed to signal an out-of-control process state, Raz [10], these are:

(i)   Outliers. A single sample beyond the OCLs (or UCL and LCL).

(ii) Trends. Consecutively increasing or decreasing sample values.

(iii) Biases. Consecutive sample values the majority of which are biased on one side or other of the central value.

(iv) Oscillations. A repetitive pattern of consecutive sample values.

Table 13.2. SPC/FAPC control strategy for process patterns (i)–(iv).

Pattern	FAPC	SPC
(i) Outliers	No	Yes
(ii) Trends	Limited	Yes
(iii) Biases	Limited	Yes
(iv) Oscillations	No	Yes

An outlier is an isolated sample value and an unsuitable candidate for FAPC which essentially takes pre-emptive action on subsequent process output, on the basis of historical evidence. SPC with production halted for investigation is the only possible course of action. Trends are caused by progressive process deterioration, environmental changes or operator fatigue for example and these may be compensated to a limited extent by FAPC. Biases that are not preceded by a trend are typically caused by a change of operator, or in an input specification, or a sudden shift in the process state or sampling system. In specific cases these may also be compensated to a limited extent by FAPC action, but more generally require external intervention. Whether oscillations can be compensated by FAPC action depends upon their periodicity and amplitude. They will be partially masked if rolling means samples are used.

Of the above control chart patterns (ii) and (iii), i.e. trends and biases are clearly capable of (limited) FAPC, whilst (i) and (iv) are most clearly SPC candidates. The proposed control strategy is summarised in Table 13.2.

Since FAPC is preferred to SPC intervention, then it is possible to consider installation of FAPC for trends and biases, provided that it can be economically justified. If trends or bias values become sufficiently large they become outliers, the process is stopped and control action transfers from FAPC to SPC. A correlated SPC/FAPC system is shown in the block diagram in Figure 13.7. The control system comprises four modules for sample analysis, signal discrimination for sorting data into FAPC or SPC processing, fuzzy control, an SPC expert function and an FTA knowledge base. The functions of these units are shown in Figure 13.7. The functions of the expert module for dispersion data are shown in some detail in Figure 13.8. This module operates with IF-THEN logic operations, for example, for dispersion data:

IF BELOW LIMIT   THEN CHECK SENSOR FUNCTIONS SENSITIVITY.

IF ABOVE LIMIT   THEN CHECK PROCESS ACCURACY COMPARED WITH LIMITS.

For central values the logic operations for a mechanical system might be:

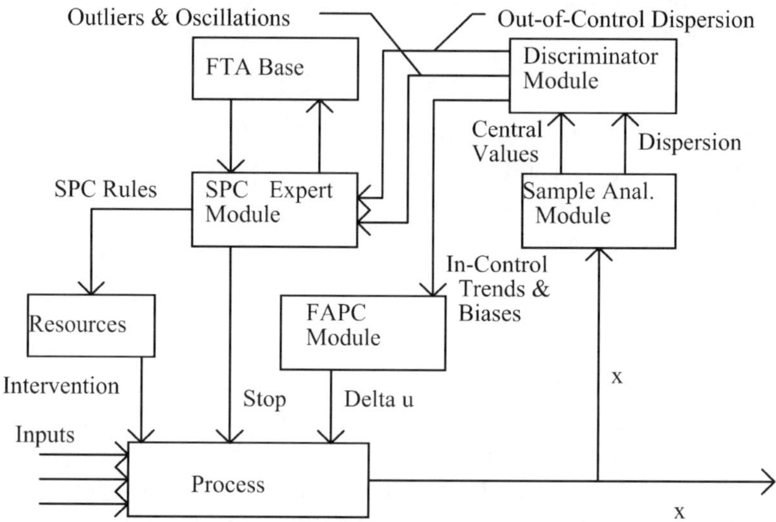

Figure 13.7. SPC/FAPC control strategy block diagram.

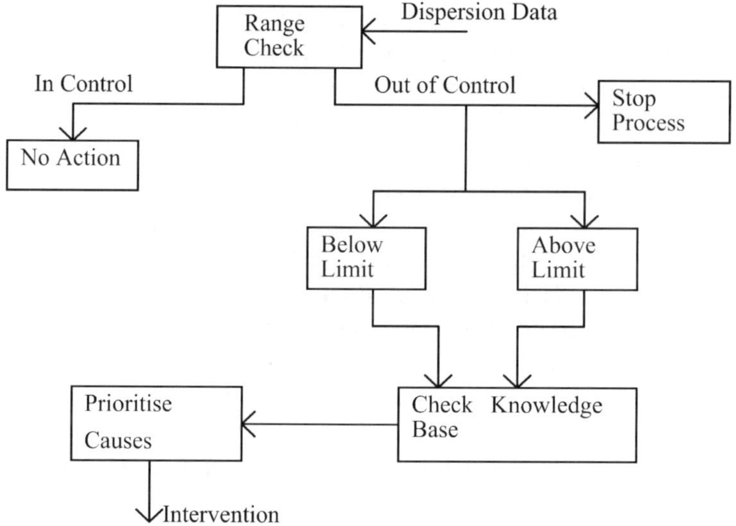

Figure 13.8. Expert module functions for dispersion data.

IF BELOW LIMIT   THEN CHECK SENSOR FUNCTIONS
IF ABOVE LIMIT   THEN CHECK   (1) PROCESS FRICTION
                                  (2) LOOSE CONNECTION
                                  (3) EXCESSIVE WEAR

where the prioritisation would be obtained from process fault tree analysis (FTA)
knowledge base.

Table 13.3. Rolling average values of sample lots without control.

Sample Lot	1	2	3	4	5	6	7	8	9	10
Aver. (mm)	5.12	4.54	4.28	4.20	4.16	4.20	3.92	3.84	4.80	5.60
Range (mm)	0.18	0.16	0.16	0.12	0.10	0.10	0.12	0.14	0.16	0.16

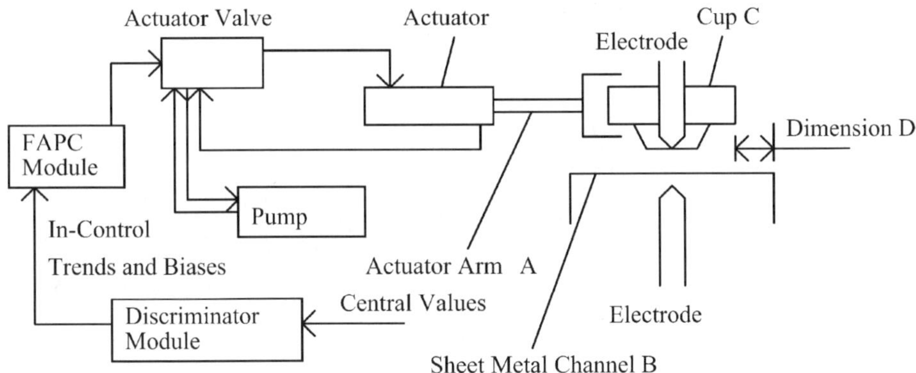

Figure 13.9. Cup/channel spot welding process.

An illustrative example of an SPC/FAPC strategy is discussed below.

EXAMPLE. Consider a simple process comprising the repetitive positioning and spot welding of a cup (C) onto a sheet metal channel (B) as shown in Figure 13.9. The cup is picked up from the storage bin and positioned over the channel by a hydraulically driven actuator arm (A). The critical parameter is dimension D, which has a target value of 5 mm. Small adjustments to the value of D are governed by FAPC through a control valve. Larger adjustments to the cup position are through the action of SPC guided maintenance intervention.

Typical observations of the rolling average and range of dimension D are shown in Table 13.3. The sample lot values are of rolling averages of five items each.

The UCL and LCL of the central values are set at 6 mm and 4 mm respectively, whilst those of the ranges are at 0.32 mm and 0.05 mm respectively. FAPC governs the central values UCL-LCL (the in-control) zone, whilst SPC governs the external (out-of-control) zone, as shown below in Figure 13.13(a).

Discussion

The central and range values listed in Table 13.3 are given in graphical form in Figures 13.10 and 13.11 respectively. From Figure 13.11 it is clear that all the range values are within limits, therefore no control action is required (only SPC intervention is associated with range control). In the case of central values,

Figure 13.10, FAPC would be initiated after sample lot 2. If the results after lot 6 were unaffected by the FAPC action, then SPC action would be initiated after lot 6. If FAPC had taken effect then lot 10 would become an outlier and SPC would be triggered at that point. The current outcome of dynamic interaction between the controls and the process would depend upon the ratio of the process time scale to that of the FAPC action and sampling time scale. It may be noted that the effect of using rolling averages and ranges is to suppress the effects of random fluctuations and oscillations on the data values.

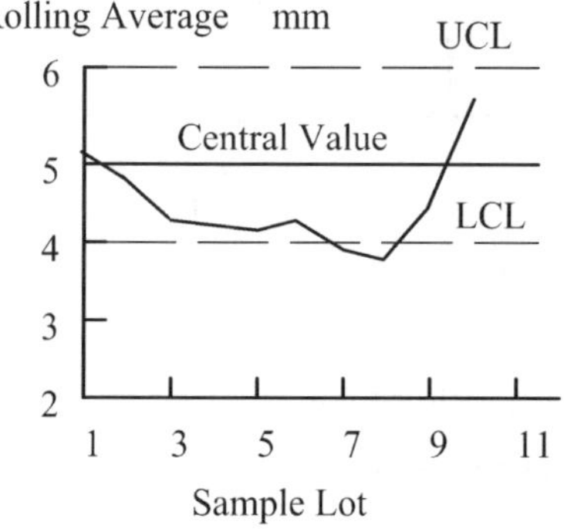

Figure 13.10. Rolling mean v. sample lot.

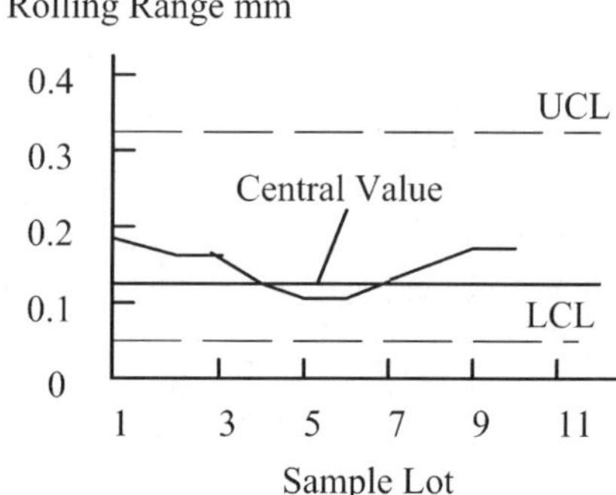

Figure 13.11. Rolling range v. sample lot.

The relationship between the actuator arm position and the control signal, u, from the FAPC module will generally be non-linear. For illustration it will be assumed to be of the type given by Equations (13.5) and (13.6), with $a = 0.5$ and $b = 0.5$. Thus,

Process governing equation:

$$z_{n+1} = 0.5 \left(\sum \delta u_i + z_n - 0.5 z_{n-1} \right), \tag{13.7}$$

where, as before,

$$z_n = 1 - x_n. \tag{13.6}$$

x_n is the current scaled value of the error in D_n. Thus,

$$x_n = (D_n - D_a)/(D_o - D_a). \tag{13.8}$$

D_o is the initial value of D and D_a is the specified target value of D_n.

For the present it will be assumed that the partitioning of the z, δz and δu spaces are as shown previously in Figures 13.4, 13.5 and 13.6 respectively. Also, that the PI controller relation at array base is given by Table 13.1.

To illustrate the FAPC action, assume in Equation (13.3) that initial values are $z_o = 0$, $z_1 = 0.2$ and $\delta z_1 = 0.2$. Membership values of fuzzy sets for these may be obtained from Figures 13.4, 13.5 and 13.6 respectively. Thus,

$$\mu_{QL1} = 0.5, \quad \mu_{QL2} = 0.5, \quad \mu_D = 1.0.$$

The fuzzy logic proposition is expressed as follows:

IF Z	AND Z	THEN U	MINIMUM	CONSEQUENCE
QL1	D	J	0.5, 1.0	0.5J
QL2	D	K	0.5, 1.0	0.5K

The fuzzy conclusion is the union of the consequences, that is,

$$\delta U_1 = 0.5J \cup 0.5K. \tag{13.9}$$

The above conclusion may be defuzzified by the application of Equation (13.10). This provides the (scaled) numerical value of the control movement,

$$\delta u_1 = \sum \mu_i \delta u_i \Big/ \sum \mu_i$$
$$= (0.5 * 0.1 + 0.5 * 0.2)/1.0 = 0.15. \tag{13.10}$$

The value of z_2 may be found from Equation (13.3), thus,

$$z_2 = 0.5(\delta u_1 + z_1 - 0.5 z_0)$$
$$= 0.5(0.15 + 0.2 - 0) = 0.175.$$

Also,

$$\delta z_2 = z_2 - z_1 = -0.025.$$

This completes the first cycle. The second cycle is similar, but in this case (and in general), the proposition has four terms as follows:

$$\mu_{QL1} = 0.563, \quad \mu_{QL2} = 0.437, \quad \mu_A = 0.25 \quad \text{and} \quad \mu_B = 0.75.$$

IF Z	AND δZ	THEN δU	MINIMUM	CONSEQUENCE
QL1	A	G	0.563, 0.25	0.25G
QL1	B	H	0.563, 0.75	0.563H
QL2	A	H	0.437, 0.25	0.25H
QL2	B	J	0.437, 0.75	0.437J

Again, the fuzzy conclusion is the union of the consequences,

$$\delta U_2 = 0.25G \cup 0.563H \cup 0.437J. \tag{13.11}$$

                    Fuzzy Logic Applications in Engineering Science

Table 13.4. Evaluation of the scaled error x.

Cycle	z_n	δz_n	δu_i	$\Sigma\delta u_i$	z_{n+1}	x_n
1	0.200	0.200	0.150	0.150	0.175	0.800
2	0.175	–0.025	0.285	0.435	0.255	0.825
3	0.255	0.080	0.224	0.659	0.413	0.745
4	0.413	0.158	0.384	1.043	0.613	0.587
5	0.613	0.200	0.103	1.146	0.776	0.387
6	0.776	0.163	0.140	1.286	0.878	0.224
7	0.878	0.102	0.128	1.414	0.952	0.122
8	0.952	0.074	0.126	1.540	1.023	0.048

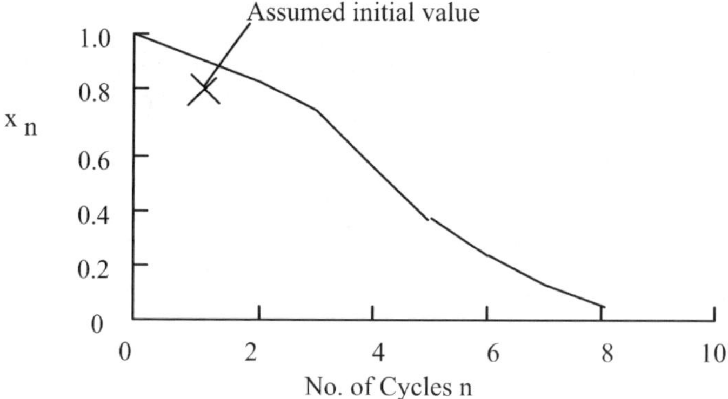

Figure 13.12. Control of error with a fuzzy PI type controller.

Defuzzifying Equation (13.11) using Equation (13.10), as before, gives a second numerical value of the control increment, $\delta u_2 = 0.285$. Using Equation (13.7) to find z_3 gives $z_3 = 0.255$. Hence, $\delta z_3 = 0.080$.

This completes the second cycle. By repeating the above process, successive values of the control increment, δu, and the corrections to the error, $x = 1 - z$, may be found. The results for eight successive cycles are shown in Table 13.4 and are displayed in Figure 13.12. After eight cycles the value of D is within 5% of its target value of 5 mm.

By revising the assumed initial value, the cycling process could be repeated to obtain improved controller input values to the process, if required.

SPC

According to the definitions and strategy adopted in this work, the SPC protocol will be activated above and below the UCL and LCL respectively as shown in Figure 13.13(a). If however the alternative OCLs and ICLs are defined, then the

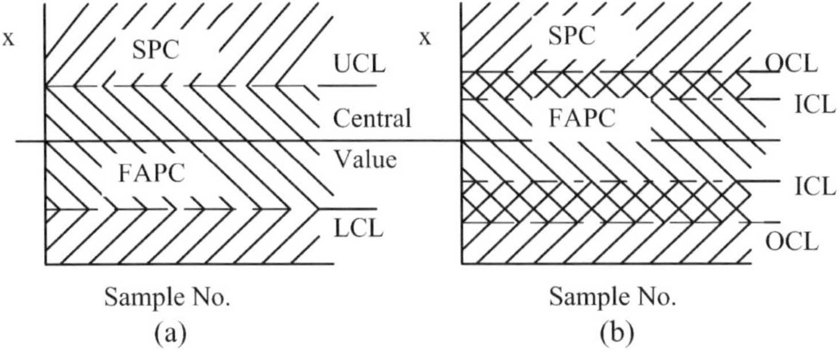

Figure 13.13. Illustration of SPC and FAPC action zones. (a) With UCL/LCL boundaries. (b) With OCL/ICL boundaries.

protocol will be activated as shown in Figure 13.13(b). In the latter case there is a soft boundary between the FAPC and SPC action zones which permits a graded engagement of the external intervention under the SPC regime as the FAPC approaches the limit of its capabilities. This can be formalised by introducing fuzzy sets to represent the FAPC and SPC zones.

SPC action is initiated by the diagnosis of sample average outliers and oscillations in the discriminator module (Figure 13.7) and by out-of-control dispersion data. Then by consultation with its library of causal rules for both sample averages and ranges, the expert module is able to advise and prioritise remedial action. The rules may be deduced from FTAs for each type of out-of-control event, such as actuator position error. FTAs are usually created during the process design stage. A simplified FTA for the actuator with sample failure probabilities in Figure 13.9 is illustrated in Figure 13.14.

The failure probabilities due to the prime faults are shown in Figure 13.14 and these enable the ordering of the most likely causes of error in the actuator position. The prime faults are shown by circles. Each of these could be taken in turn as a top event for further FTA taking the analysis to greater levels of detail. From Figure 13.14, the likelihood of position error may have the following causes, in order of probabilities:

(i)   Pump electric fault (EL),
(ii) Pump mechanical fault (ME),
(iii) Actuator valve fault (VA),
(iv) Hydraulic fluid leak (HY).

The above list also gives the preferred order of investigation for corrective maintenance. The order would be subject to revision in the light of further operating experience. Rules for the causes of errors in the observed averages of the critical parameter (in this case, dimension D in Figure 13.9) are summarised in Table 13.5.

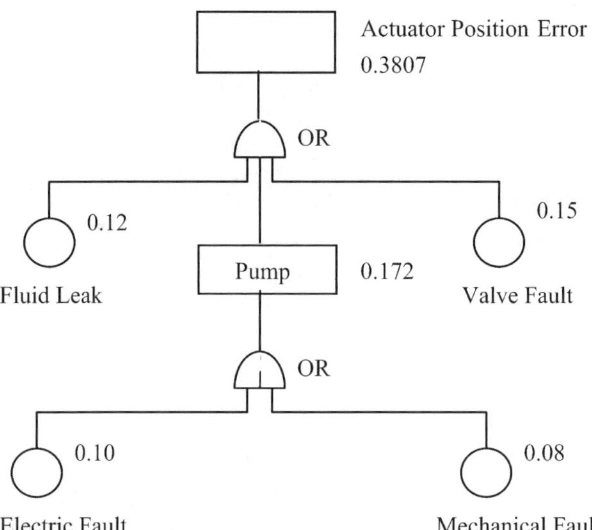

Figure 13.14. Simplified fault tree for the actuator average position error (top event).

Table 13.5. Summary of cause and effect rules for averages.

	Effect			
Cause	**OU**	**TR**	**BI**	**OS**
EL	#		#	#
ME	#	#	#	
VA	#		#	#
HY	#	#	#	

OU = outlier, TR = trend, BI = bias,
OS = oscillations.

The dispersion causes are usually more fundamental, as discussed earlier. Combinations of the effects in Table 13.5 are possible. This type of information provides the FTA knowledge base for the expert module, as shown in Figure 13.7. The expert module functions, such as those shown in Figure 13.8 for example, are the basis of logical propositions of the form

IF A   AND B   THEN C_1 OR C_2 OR C_3 OR C_4 etc.

EXAMPLE. Consider again the simple manufacturing process shown in Figure 13.9 and the sample lot data shown in Figure 13.10, but in this case attention is focused upon the two outliers provided by sample lots 7 and 8. Assuming the ICL/OCL type of pattern shown in Figure 13.13(b).

Discussion

If the earlier FAPC action has not changed the outlier nature of sample lots 7 and 8, they could be classed as "outlier plus bias". From Table 13.5, the causes are one or more of (EL, ME, VA, HY) in that order of probability. Corrective maintenance checks in the same order would be advised by the expert module. If however, the control strategy adopted is that illustrated in Figure 13.13(b), then some of the sample values may fall within the FAPC/SPC overlap zone. This is the zone of fading FAPC effectiveness and the expert module would advise SPC alert, visual process check without process shut-down, moving on to process "stop" and corrective maintenance if sample data enters the SPC only zone.

13.5. Conclusion

During the commissioning phase of a process it is necessary to establish the normal average value of a given critical parameter. In some cases this is specified as a design target value. In this phase a pre-production batch of some fifty to perhaps several hundred samples will be checked after process adjustments to determine process performance and an initial statistical profile of the products.

In the mature production phase, the critical parameter raw data includes "noise" due to random events (including those within the sensing system) and oscillations which are superimposed on the average values. Averaging by taking samples of say, five consecutive products will provide some reduction of these effects. Their effect is further reduced by finding rolling averages of say, five or more consecutive samples to provide preliminary data somoothing.

For the control function the UCLs (or OCLs) are often located at an estimated three standard deviations of the universe from the central value, this gives a 99.8% sample inclusion based upon a Gaussian distribution of the population. The location of the control limits (including the ICLs) is of course discretional, they may be located at any selected point relative to the central value. By convention, sample values beyond the UCL or LCL (or OCLs) are deemed to signal an "out-of-control" process state and a requirement for off-line corrective intervention and the same convention is adopted in this work. Within the UCL-LCL (or OCLs) boundaries the process is subject to FAPC. Such a control strategy preserves the benefits of each type and provides optimum control in terms of orocess plant productivity.

The problem of sampling and interpretation of the data is clearly complex and needs careful consideration. Process control sampling and consequent control action should be undertaken as near to the critical parameter production point as possible to ensure the most effective and economical process operation. The control response time, in terms of the number of production units, is then a minimum for a given control strategy.

13.6. Notation

a, b	coefficients in Equation (13.5)
i	ith fuzzy set, or non-current cycle number
n	current cycle number, or sample number
r_a	central value of dispersions
r_i	inner control limit of dispersions
r_o	outer limit of dispersions
u_i	ith controller input
x	sample average value
x_a	central value of averages
x_o	outer control limit of averages
x_i	inner control limit of averages
z_n	nth controller input
$A(n), B(n), C(n), D(n)$	tabulated coefficients
U	controller output fuzzy set
Z	controller input fuzzy set
δ	increment of …
μ_i	membership value of the ith fuzzy set
σ	standard deviation of the universe.
$\cup$	logic union

References

1. Raadnui, S., "Condition Monitoring of Earth Moving Equipment – A Novel Approach", in *3rd Intnl. Conf. on Quality, Relaibility and Maintenance*, G.J. McNulty (ed.), Prof. Engng. Publ. Ltd., Suffolk, UK, 2000, pp. 159–162, ISBN 1-8605-8256-7.
2. Kreysig, E., *Advanced Engineering Mathematics*, 3rd edn, Wiley and Sons Inc., New York/London 1972.
3. Harris, J., *An Introduction to Fuzzy Logic Applications*, Chapter 5, Kluwer Academic Publishers, Dordrecht, 2000, ISBN 0-7923-6325-6.
4. Rowlands, H. and Wang, L.R., "An Approach of Fuzzy Logic Evaluation and Control in SPC", *Qual. and Relaib. Engng. Intnl.*, Vol. 16, 2000, pp. 91–98.
5. Wang, L.R. and Rowlands, H., "A Fuzzy Logic Application in SPC Evaluation and Control", in *7th IEEE Intnl. Conf. on Emerging Tech. and Factory Automation*, Barcelona, Spain, pp. 679–684.
6. Davies, O.L., *Statistical Methods in Production and Research*, Oliver and Boyd, London, 1958.
7. Thompson, S., *Control Systems Engineering and Design*, Longman Scientific and Technical, Harlow, 1989, ISBN 0-5829-9468-3.
8. Reznik, L., *Fuzzy controllers*, Newnes, Oxford, 1997, ISBN 0-7506-3429-4.
9. Davidson, J. (ed.), *The Reliability of Mechanical Systems*, 2nd edn, Chapter 13, Mech. Engng. Publ. Ltd., London, 1994, ISBN 0-8529-8881-8.

10. Raz, T., "Control Chart", in *McGraw-Hill Encycl. of Science and Tech.*, Vol. 4, McGraw-Hill, London/New York, 1997, pp. 405–410, ISBN 0-0791-1504-7.
11. Raz, T., "Quality Control", in *McGraw-Hill Encycl. of Science and Tech.*, Vol. 14, McGraw-Hill, London/New York, 1997, pp. 645–649, ISBN 0-0791-1504-7.

Chapter 14
Road Transport Fuzzy Logistics – A Case Study

Keywords. Zoning, Array, Groups, Similitude, Composition, Fuzzy logic.

Abstract. This work reports on an initial investigation for a management study of a small to medium sized road transport service. The service offers a limited collection and delivery service for small and large consignments. The geographical range of about 160 km radius is divided into seven zones of equal area for the study. Using records of recent years, the service demands of each zone are assessed and tabulated. This data bank is used to identify similar adjacent zones and also for the analysis of linkage strengths between collection and delivery groups. It provides a rational basis for a higher level of operational matching of demand with physical and manpower resources and also for a more detailed financial management control.

14.1. Introduction

Small and medium sized enterprises (SMEs) have strengths in local knowledge and contacts, in personal service and also in flexibility and adaptability. They are weakest in economies of scale and in volume trading advantages which confer economic strengths to large organisations, which are also able to sustain more powerful advertising campaigns. One of the particularly damaging factors in the current operations of SME transport companies is that of rising fuel costs and this pressure is likely to increase in the foreseeable future. Whilst vehicle manufacturers will continue to search for more efficient engine formats and also for alternatives to the ubiquitous liquid fossil fuel energy sources, these are probably longer term solutions. In the meantime, SME transport companies must seek ways of improving their operating efficiencies as one way of ensuring their continued survival in the short to medium term. It is not only profitability that is of interest, but also sustainability, reputation and growth that are important. On a broader front, this also leads to reduced pollution, which is one point where commercial and environmental interests converge. This work is part of a business development project to improve competitiveness by maximising the effective use of resources.

Developments in the theory and practice of manufacturing technology have been very rapid and far reaching during the recent past. Amongst the more significant paradigms emerging have been the so-called soft computing methods, amongst which are those of fuzzy logic, artificial neural networks and genetic algorithms.

These enable studies to be conducted in an entirely different framework to before. The present work is concerned with the application of similitude to a management study on an SME road transport service, the antecedents of which are to be found in the manufacturing technology sector, see for example Harris [1]. To date there are few publications of fuzzy logic applications available in the service industries sector, though clearly there is considerable scope for this. Background fuzzy logic theory and applications may be found in the texts by Ross [2] and by Harris [3]. In applications of operational research, the transportation problem is treated as a branch of linear programming in which special methods are used, because in the normal linear programming methodology the number of decision variables, which in general is the product of the number of sources and destination, and also the number of constraints, which equals the sum of the number of sources and constraints, often becomes unwieldy, as described by Lesso [4]. A basic heuristic is the so-called North-West Corner rule for distributing limited numbers of undifferentiated units from a finite number of sources to a finite number of destinations. VogelŠs approximation method provides an alternative treatment. The assignment problem is another special form of transportation problem in which an unlimited number of differentiated units from a finite number of sources are assigned to a finite number of destinations on a one-to-one basis. This may be solved by the stepping stone or the Hungarian method as described, for example, by Urry [4]. (It may also be solved by a similitude method, similar to the one described here, but is not treated in this work). In the present work the transfer units are differentiated and addressed, and in general sources and destinations are linked in a multiplicity of ways. Moreover (in one case) there is a sorting operation.

14.2. Organisation Outline

The present operations of the Company under study, which would rank as an SME, are based upon conventional districts surrounding the depot. The topography is divided into a series of zones of increasing radius from the depot to the outer limit of operations, which is about 160 km radius. Radial distances previously provided a basic distance reference to be used in conjunction with a scaling parameter for tariffs, but there are now software packages which facilitated journey distance estimation and which are the main pricing tool.

The Company currently has a fleet of 28 vehicles, some suitable for light work and some for heavier duty. About 24 vehicles are operational at any one time, the non-functional part of the fleet being on maintenance or without drivers. Shortfalls can be offset by vehicle hire, provision is also made for surplus capacity to be hired out. Vehicle running time averages 6 hours per day, giving a road distance capacity of about 288 km/day or 48 km/hour/vehicle. The fleet capacity is therefore 6,912 km/day. The target operation is 65% of capacity, that is, 4,493 km/day. The average payload is 75 tonnes, the target fleet load is therefore 3,370 km tonnes/day.

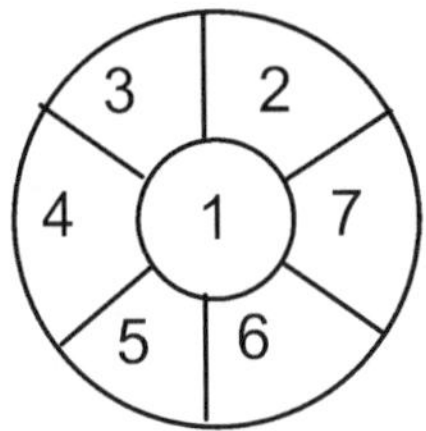

Figure 14.1. Zoning pattern of the operating area.

The break-even point is estimated as 60% of the target load, that is 2,022 km tonnes/day.

The fleet operations are organised on a daily basis by the fleet manager, based upon prior experience. Drivers usually prefer to operate regular districts and part load aggregation is largely on an ad hoc basis, as is the general matching of demand and resources.

Tariffs are based upon a two tier system. Tier 1 applies to smaller items up to 40 kg for individual items that have a girth plus length not exceeding 3.5 metres. Tier 2 applies to larger consignments of up to 250 kg for individual items. These two tariff tiers are maintained in the following study.

14.3. Analysis of Operations

14.3.1. BASIC DATA

The basic data comprises records of collection and delivery points and also the type and size of consignment for each of the two tiers mentioned above. Records were available for about ten years, but only the more recent records were relevant due to changes in the road system and other developments. The records were recast in terms of a grid of seven zones comprising; central, east, northeast, northwest, west, southwest and southeast zones. These zones were numbered 1–7, as shown in Figure 14.1.

The zones in Figure 14.1 are chosen to have equal areas, thus,

$$a_1^2 = (a_2^2 - a_1^2)/6. \tag{14.1}$$

Hence,

$$a_2 = 7^{1/2}a_1, \tag{14.2}$$

where a_1 is the radius of the inner (central) zone and a_2 is the outer radius. Since a_2 is 160 km, then a_1 is about 60 km.

Tables showing loading according to a weight/size formula of collection and delivery zones, scaled to the maximum values, are shown below. Table 14.1 shows the results for small consignments, whilst the results for large consignments are shown in Table 14.2. The horizontal rows in Tables 14.1 and 14.2 may be viewed as fuzzy

Table 14.1. Tier 1. Small consignment relative loading.

| Collection | Delivery Zone | | | | | | | |
Zone	1	2	3	4	5	6	7	Total
1	0.224	0.630	0.597	0.046	0.099	0	0.099	1.695
2	0.450	0.531	0.458	0.024	0.812	0.694	0.471	3.440
3				Negligible				0
4	0.192	0.208	0.721	0.340	0.106	1.000	0.436	3.003
5	0.022	0.250	0.258	0.892	0.634	0.824	0.317	3.197
6	0.191	0.880	0.699	0.447	0.965	0.522	0	3.704
7	0	0.142	0	0.011	0.614	0.550	0.085	1.402

Table 14.2. Tier 2. Large consignment relative loading.

| Collection | Delivery Zone | | | | | | | |
Zone	1	2	3	4	5	6	7	Total
1	0.573	0.359	0.276	0.669	0.401	0	0.603	2.881
2	0.016	0.582	0.014	0.031	0.019	0.931	0.063	1.656
3				Negligible				
4	0.860	1.000	0.619	0.692	0.587	0.575	0.463	4.796
5	0.529	0.456	0.842	0.411	0.618	0.751	0.608	4.215
6	0.552	0.210	0.636	0.566	0	0.596	0.194	2.754
7				Negligible				

sets on the delivery zone universal of discourse. Correspondingly, the columns in these tables may be viewed as fuzzy sets on the collection zones universe of discourse.

In these tables, a value of zero or negligible means <0.001. The tabulated values are scaled to the maximum at $\{4, 6\}$ in Table 14.1 and $\{4, 2\}$ in Table 14.2. The total of each collection zone value is shown in the right-hand column, the maximum is for Collection Zone 6 in Table 14.1 and Collection Zone 4 in Table 14.2. The maximum tabulated value is for Delivery Zone 6 in Table 14.1 and it originates from zone 4, whilst in Table 14.2 the maximum in Delivery Zone 6 originates from zone 2.

Other features may be noted from Tables 14.1 and 14.2, but rationalisation of the collection and delivery operations would not be easy by inspection and therefore a systematic analysis of the data was undertaken as outlined below.

14.3.2. SMALL CONSIGNMENT ANALYSIS

In the case of small consignments, delivery is normally made to the depot by the consignor, but collection could be arranged. In either case, consolidation of the consignments is sought each day on the basis of zoning shown in Figure 14.1, aimed at maximising vehicle utilisation and other benefits described later. However, further information was sought on possible favourable combinations of zones as an additional means of aggregating loads, both for collection and delivery.

As a first step, it was considered that both collection and delivery utilisation of resources could be improved for part loads by pairing adjacent zones. The most effective method of achieving this would be by pairing of the most similar zones with respect to both collection and delivery. Therefore a similarity analysis of the data in Tables 14.1 and 14.2 was conducted by columns and rows to provide a basis for selection.

There are various methods of conducting similarity analysis, such as the min/max, cosine amplitude, correlation coefficient and others, as described by Ross [2]. All the methods should give corresponding results, though numerical values will differ. The min/max method (called max-min by Ross [2]) was chosen in this case as being computationally simple, but effective.

Consider similarity of two rows, i, k, of a table, such as Table 14.1. The similarity metric rik is defined by

$$r_{ik} = \min(x_{ij}, x_{kj})/\max(x_{ij}, x_{kj}) \tag{14.3}$$

(summation over j: $j = 1, \ldots, n$).

A similar operation is carried out on column pairs. Thus, considering columns 3 and 6 of Table 14.1, the similarity metric is given by

$$
\begin{aligned}
r_{36} = & [\min(0.597, 0) + \min(0.458, 0.694) + \min(0, 0) + \min(0.721, 1.000) + \\
& + \min(0.258, 0.824) + (0.699, 0.522) + \min(0, 0.550)]/ \\
& [\max(0.597, 0) + \max(0.458, 0.694) + \max(0, 0) + \max(0.721, 1.00) + \\
& + \max(0.258, 0.824) + (0.699, 0.522) + \max(0, 0.550)] \\
= & \ 0.449 \text{ etc.}
\end{aligned}
$$

and obviously, $r_{36} = r_{63}$. Also $r_{ii} = 1$ for all i. The resulting array is therefore symmetrical and the leading diagonal entries are all unity. The similarity metric arrays for columns (delivery zones) and rows (collection zones) are given in Tables 14.3 and 14.4 respectively.

At this stage, similitude may be sought from Tables 14.3 and 14.4. This is accomplished by considering descending levels of cut-off (λ), commencing with the highest value of r_{ik} and proceeding stepwise to successively lower levels of λ, matching lower values of rik. Consider Table 14.3 with $= 0.597$ (i.e., r_{56}), the cell pattern in Table 14.5 is obtained.

Table 14.3. Delivery zones similitude array.

		1	2	3	4	5	6	7
	1	1	0.409	0.395	0.201	0.252	0.226	0.443
	2		1	0.461	0.282	0.519	0.361	0.421
	3			1	0.330	0.373	0.449	0.433
$R =$	4				1	0.355	0.444	0.303
	5		Sym			1	0.597	0.305
	6						1	0.340
	7							1

Table 14.4. Collection zones similitude array.

		1	2	3	4	5	6	7
	1	1	0.388	0	0.359	0.188	0.497	0.122
	2		1	0	0.490	0.495	0.551	0.408
	3			1	0	0	0	0
$S =$	4				1	0.503	0.445	0.255
	5		Sym			1	0.447	0.439
	6						1	0.338
	7							1

Table 14.5. Similitude pattern for $\lambda \geq 0.597$ for R.

	1	2	3	4	5	6	7
1	1						
2		1					
3			1				
4				1			
5					1	1	
6					1	1	
7							1

Table 14.6. Similitude pattern for $\lambda \geq 0.519$ for R.

	1	2	3	4	5	6	7
1	1						
2		1		1			
3			1				
4				1			
5		1			1	1	
6					1	1	
7							1

It may be noted from Table 14.5 that columns (and rows) $\{5, 6\}$ are similar.

Next, the entries in Table 14.3 are selected with $\lambda = 0.519$ or above, the corresponding cell pattern is shown in Table 14.6.

It may be noted that no similarities exist at this cut-off level. Proceeding through successively lower cut-off levels, the next similarity pattern is found at $\lambda = 0.395$ (i.e. r_{17}), which means that columns (and rows) $\{1, 7\}$ are similar. Then at $\lambda = 0.361$, etc. A summary of the results is listed below.

λ	Delivery Groups	λ	Delivery Groups
0.597	$\{5, 6\}$	0.433	–
0.519	–	0.421	–
0.461	–	0.409	–
0.449	–	0.395	$\{1, 7\}$
0.444	–	0.373	$\{1, 7\}$
0.443	$\{1, 7\}$ 0.361	$\{1, 7\}\{2, 3\}$	

This exhausts the possible similarity groups. Zone $\{4\}$ remains to form its own group.

The order of preference of adjacent zones is: $\{5, 6\}$, $\{1, 7\}$, $\{2, 3\}$ and $\{4\}$.

By the same procedure the S array in Table 14.4 may also be searched for similitude of the collection zones. This results in the following list of groups for descending cut-off levels:

λ	Collection Groups	λ	Collection Groups
0.551	{2, 6}	0.408	{2, 5}
0.503	{4, 5}	0.388	–
0.497	{4, 5}	0.359	{4, 6}
0.495	{4, 5}	0.338	{2, 6}
0.490	{4, 5}	0.255	{2, 4, 6}{5, 7}
0.447	{2, 5}	0.188	{2, 4, 5, 6}
0.445	{2, 4, 5}	0.122	{1, 2, 4, 5, 6, 7}
0.439	{2, 4}		

There are no collections from Zone 3.

According to the above list, the most favourable adjacent pairs are {4, 5}, whilst {1, 2} and {6, 7} only appear together at the cut-off level, $\lambda = 0.122$. The order of preference of groups is: {4, 5} and {1, 2}, {6, 7}, the latter two being of equal preference.

Similarity can however be intensified by a composition operation on each of the above arrays. This operation drives the similitude arrays from a tolerance condition, Ross [2], towards an equivalence condition. Composition is defined by

$$R^2 = R \circ R \tag{14.4}$$

and

$$S^2 = S \circ S. \tag{14.5}$$

For example,

$$\begin{aligned}
r_{46}^2 &= \max[\min(0.201, 0.226), \min(0.282, 0.361), \min(0.330, 0.449), \\
&\quad \min(1.000, 0.444), \min(0.355, 0.597), \min(0.444, 1.000), \\
&\quad \min(0.303, 0.340)] \\
&= 0.444 \text{ etc.}
\end{aligned}$$

With a similar operation to find the S^2 array. The resulting arrays are shown in Tables 14.7 and 14.8.

Table 14.7. Small load R^2 array.

		1	2	3	4	5	6	7
	1	1	0.421	0.433	0.330	0.409	0.395	0.443
	2		1	0.461	0.361	0.519	0.519	0.433
	3			1	0.438	0.461	0.449	0.433
$R^2 =$	4				1	0.438	0.444	0.340
	5		Sym			1	0.597	0.421
	6						1	0.433
	7							1

Table 14.8. Small load S^2 array.

		1	2	3	4	5	6	7
	1	1	0.407	0	0.407	0.407	0.407	0.388
	2		1	0	0.495	0.495	0.551	0.439
	3			1	0	0	0	0
$S^2 =$	4				1	0.503	0.490	0.439
	5		Sym			1	0.495	0.439
	6						1	0.439
	7						1	

The above arrays are now searched for similitude, following the same method as previously. The results of the search for delivery group similarities is listed below.

λ	Delivery Groups	λ	Delivery Groups
0.597	{5, 6}	0.449	{2, 3, 5, 6}
0.519	{2, 5, 6}	0.444	{2, 3, 5}
0.461	{2, 5}	0.443	{1, 7}{2, 3, 5}

The order of preference in this case is: {5, 6}, {2, 3}, {1, 7} and {4}, which is achieved in 6 steps in this case, compared with 12 steps required for the R array above.

The results obtained for the collection group similarities is listed below.

λ	Collection Groups	λ	Collection Groups
0.551	{2, 6}	0.439	{2, 4, 5, 6, 7}
0.503	{2, 6}{4, 5}	0.407	{2, 4, 5, 6}
0.495	{2, 5}	0.388	{1, 2, 4, 5, 6, 7}
0.490	{2, 4, 5, 6}		

The order of preference in this case is: {4, 5}, {6, 7}, and {1, 2}. Seven steps are required here compared with 15 for the S array above. This illustrates the higher resolving power obtained by the introduction of the composition operation after the initial similitude operation.

The forward linkage strengths between the Collection Zone Groups and the Delivery Zone Groups express the relationship array between Collection and Delivery Groups. The right-hand column in Table 14.1 shows the total relative load for each Collection Zone. The linkage strength is defined as the fraction that is destined for the respective Delivery Zones. Thus, considering the pairs already identified, for the Collection Zone Group {6, 7} and the Delivery Zone Group {2, 3}, the linkage strength (w) is

$$w_{\{6,7\}\{2,3\}} = (r_{62} + r_{63} + r_{72} + r_{73}) \bigg/ \sum (r_{6k} + r_{7k}), \tag{14.6}$$

where the summation is carried out between the limits $1 \leq k \leq 7$.

$$w_{\{6,7\}\{2,3\}} = (0.880 + 0.142 + 0.699 + 0)/(3.704 + 1.402)$$

$$= 0.337 \text{ etc.}$$

The complete array of linkage strengths for small consignment collection and delivery is shown in Table 14.9.

From this table it may be noted that the strongest linkages are as follows:

Collection Zones	Delivery Zones
1, 2	2, 3
4, 5	5, 6
6, 7	5, 6

Whilst for comparison, the weakest linkages are:

Collection Zones	Delivery Zones
1, 2	4
4, 5	1, 7
6, 7	1, 7

Table 14.9. Small consignment linkage strengths.

Collection Zone Group	Delivery Zone Group				
	5, 6	**2, 3**	**1, 7**	**4**	**Row Total**
1, 2	0.313	0.432	0.242	0.014	1.001
4, 5	0.414	0.232	0.156	0.199	1.001
6, 7	0.519	0.337	0.054	0.090	1.000
Total	1.246	1.001	0.452	0.303	

Table 14.10. Large consignment R^2 array.

		1	2	3	4	5	6	7
	1	1	0.587	0.680	0.845	0.661	0.586	0.661
	2		1	0.587	0.587	0.586	0.501	0.587
	3			1	0.680	0.594	0.586	0.661
$R^2 =$	4				1	0.655	0.586	0.661
	5		Sym			1	0.586	0.722
	6						1	0.563
	7							1

14.3.3. LARGE CONSIGNMENT ANALYSIS

There are similarities between large and small consignment analyses, but there is an operational difference in that large consignments are usually delivered direct without passing through the sorting process. There are also fewer opportunities for part-load aggregation in the field.

The similarity analysis may proceed as illustrated above in Section 14.3.2. Thus, the results of composition operations are shown below in Tables 14.10 and 14.11.

Table 14.11. Large consignment S^2 array.

$$S^2 = $$

	1	2	3	4	5	6	7
1	1	0.571	0	0.571	0.571	0.571	0
2		1	0	0.571	0.571	0.571	0
3			1	0	0	0	0
4				1	0.682	0.586	0
5		Sym			1	0.586	0
6						1	0
7							1

As before, the above R and S arrays are now searched for similarities with descending cut-off levels. The respective lists of zonal groupings are shown below:

λ	Delivery Groups	λ	Delivery Groups
0.845	$\{1, 4\}$	0.594	$\{1, 3, 4, 5, 7\}$
0.722	$\{1, 4\}\{5, 7\}$	0.587	$\{1, 3, 4, 7\}$
0.680	$\{1, 3, 4\}\{5, 7\}$	0.586	$\{1, 2, 3, 4, 5\}$
0.661	$\{1, 7\}\{3, 4\}$	0.563	$\{1, 3, 4, 5, 6, 7\}$
0.655	$\{1, 4, 7\}$		

The order of preference of the zonal groups is: $\{1, 4\}$, $\{2, 3\}$ and $\{5, 6, 7\}$. Zones 5, 6 and 7 may be paired as $\{5\}\{6, 7\}$ or $\{5, 6\}\{7\}$.

Now considering the collection groups:

λ	Collection Groups
0.682	$\{4, 5\}$
0.586	$\{4, 5, 6\}$
0.571	$\{1, 2, 4, 5, 6\}$

The order of preference of these groups is: $\{4, 5\}$ and $\{1, 2\}$ and $\{6\}$. The latter groups are of equal order.

In this case, the array of linkage strengths for large consignment collection and delivery zones is shown in Table 14.12.

Table 14.12. Large consignment linkage strengths.

Collection Zone Group	Delivery Zone Group				
	1, 4	**5, 6**	**2, 3**	**7**	**Row Total**
1, 2	0.284	0.298	0.271	1.147	1.000
4, 5	0.277	0.281	0.324	0.119	1.001
6	0.406	0.216	0.307	0.070	0.999
Total	0.967	0.795	0.902	0.336	3.000

From this table it may be noted that the strongest linkages are as follows:

Collection Zones	Delivery Zones
1, 2	5, 6
4, 5	2, 3
6	1, 4

Whilst for comparison, the weakest linkages are:

Collection Zones	Delivery Zones
1, 2	7
4, 5	7
6	7

14.4. Organisational Effects

A property of the system outlined in this work is that it embodies intelligence based upon records of prior operations. It is obvious that a judicious combination of zones can be chosen for both part-load collection and part-load delivery, rather than an ad hoc selection.

The linkage strengths shown as the relational array in Table 14.9 can be translated into a best sorting operation layout, as shown in Figure 14.2.

There is a rough sorting process by maintaining the distinction between Collection Zones {1, 2}, {4, 5} and {6, 7}. By the array in Table 14.9, they can be linked to the most likely destination. Zone 4 is the most unlikely destination, and can therefore be located most remotely, but accessible to, the receiving points shown in Figure 14.2. The collection zone receiving points can be associated closest to

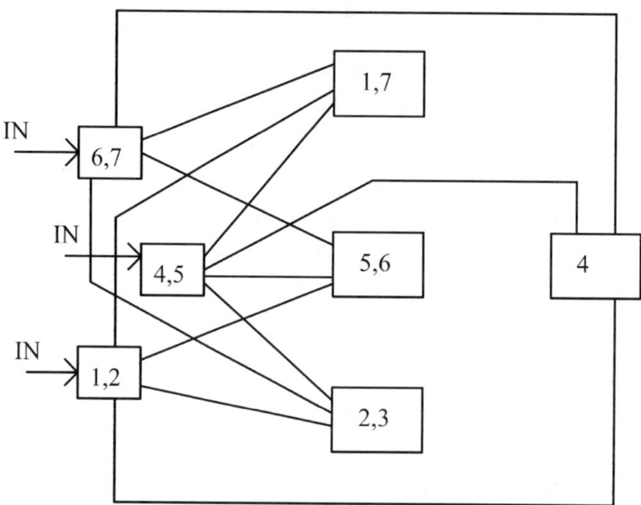

Figure 14.2. Sorting operation small consignment floor layout.

Table 14.13. Small consignment aggregate relative loads.

Collection Zone Group	Delivery Zone Group				
	2, 3	**5, 6**	**1, 7**	**4**	**Total**
1, 2	2.216	1.605	1.244	0.070	5.135
4, 5	1.437	2.564	0.967	1.232	6.200
6, 7	1.721	2.651	0.276	0.458	5.106

the destinations with which they have the strongest links to make an ergonomically sound floor plan.

To balance resources and demands, the allocation of physical, financial and manpower resources should be in proportion to the relative aggregate loading of the various linkages that have been established in the above analysis. The data in Tables 14.1 and 14.9 may be presented in aggregate form as shown in Table 14.13.

Comparing the heaviest with the most lightly loaded route, {6, 7}/{5, 6} route is about 38 times that of the {1, 2}/{4} route. The allocation of physical, financial and manpower resources should be allocated in proportion to the relative loading shown in Table 14.13.

14.4.1. LARGE CONSIGNMENTS

In the case of large consignments, the sorting process is absent, though there may be occasional load aggregation in the field, but not as frequently as in the case of small consignments. The group loading pattern can influence the allocation of

Table 14.14. Large consignment group relative loads.

Collection Zone Group	Delivery Zone Group				
	1, 4	**2, 3**	**5, 6**	**7**	**Total**
4, 5	2.492	2.917	2.531	1.071	9.011
1, 2	1.289	1.231	1.351	0.666	4.537
6	1.118	0.846	0.596	0.194	2.754

resources, as is the case for small consignments. The large consignment group relative loading pattern is shown in Table 14.14.

Table 14.14 illustrates the relative market demand by zonal groups. Collection Zone Groups {4, 5} clearly makes the major demand on operational resources and delivery to Group {2, 3} takes the largest share. The least demand is from Collection Zone Group {6} and of this the least share is to Group {7} Delivery. A scaling of this type is obviously useful when transferring resources up or down the demand pattern to balance the overall load matching, as would also be the case for small consignments.

The pattern in Table 14.14 (and also in Table 14.13) has finical accounting implications, in that each Collection/Delivery Zone Group can be viewed as a cost centre and the nett profitability of the twelve centres can be evaluated and compared over a given period. On the administrative side, staff responsibility can be allocated to groups in proportion to the market demand pattern and the costs apportioned to each cost centre to be included in the evaluation of the nett profitability of each centre. This enables a refined financial analysis to be conducted to identify the least profitable centres.

14.5. Conclusions

Data records of a transport company have been reorganised on a cellular zonal basis with seven zones of equal area. The records of small and large consignment movements are then analysed on a similarity basis, backed up by composition which can reduce the searching for similarity. This results in zonal pairs displaying the greatest similarity. The operation has been applied to both Collection and Delivery Zones and enables linkage strengths between the two types of zone to be evaluated. The relative service demands by the Collection Zone Groups for small consignments is portrayed in Table 14.13, whilst that for large consignments is shown in Table 14.14. The management benefits of the analysis are:

1. Efficient aggregation of part loads,
2. An enhanced floor layout for small consignment sorting,
3. A basis for apportioning of physical resources for load matching,

4. A basis for the allocation of financial resources,
5. Scaling of the allocation of available operational and administrative manpower,
6. The identification of cost centres for comparative profitability analysis,
7. The more formalised presented here renders the business less vulnerable to the effects of management staff changes.

This methodology extracts the maximum benefit from the data records and provides a more intelligent system than the alternative non-cellular zones and less formalised structure. Finer detail could be implemented by reducing the grid size of the zones in Figure 14.1, which would mean increasing the number of zones. This would increase the numerical work, but there is no problem about computerising the process.

It is likely that when fully implemented, the system will be based upon a grid of nineteen zones of equal area comprising a central zone surrounded by a ring of 6 zones and an outer ring of 12 zones. The radial boundaries being at 37, 97 and 160 km respectively.

14.6. Notation

λ cut-off level

a_1 inner zone radius
a_2 outer zone radius
r_{ik} similarity metric
x_{ik} array entry
w linkage strength

R delivery zone array
S collection zone array

References

1. Harris, J. and Kanhukamwe, Q.C., "Similarity Treatments of Group Technology", in *Proc. 3rd Intn. Conf. on Manuf. Processes, Systems & Opns. Man. in LIRs*, Nat. Univ. Sci. & Tech., Bulawayo, Zimbabwe, 1998, pp. 62–67, ISBN 0-7974-1794-X.
2. Ross, T.J., *Fuzzy Logic with Engineering Applications*, McGraw-Hill Inc., 1995, ISBN 0-0711-3637-1.
3. Harris, J., *An Introduction to Fuzzy Logic Applications*, Kluwer Academic Publishers, 2000, ISBN 0-0791-1504-7.
4. Urry, S., *Introduction to Operational Research*, Longman Scientific & Technical, 1991, ISBN 0 5820-1349-6.

Further Reading

The following are entry point references for background reading. A comprehensive and frequently updated list of the latest research publications in the field may be found on the website: www.evl.ac.uk/ram/ search for Fuzzy Logic. A general review of fuzzy logic and applications is given in *Fuzzy Logic with Engineering Applications*, by T.J. Ross, published by McGraw-Hill Inc., USA, 1995, ISBN 0-7113-6371. A wide range of engineering and management applications may be found in *An Introduction to Fuzzy Logic Applications*, by J. Harris, published by Kluwer Academic Publishers, The Netherlands, 2000, ISBN 0-7923-6325-6.

Chapter 1
The above two texts are relevant.

Chapter 2
Tolerance of form, orientation, location and run-out. BS ISO 1101:1983.
Geometric Product Specifications. BS ISO 4288:1996.

Chapter 3
Hopgood, A.A., *Knowledge Based Systems for Engineers and Scientists*, CRC Press, 1993, ISBN 0-8493-8616-0.
Ashby, M.F., "On the Engineering Properties of Materials", *Acta Metallurgica*, Vol. 37, 1989, P1273.

Chapter 4
Hunsaker, J.C. and Rightmire, B.G., *Engineering Applications of Fluid Mechanics*, McGraw-Hill Inc., USA, 1947.
Dowson, D. and Miranda., A.A.S., "The Prediction of Liquid Film Journal Bearing Performance with a Consideration of Lubricant Film Reformation. Part 1, Theoretical Results", *Proc. Instn. Mech. Engrs.*, Vol. 199, No. C2, 1985, pp. 95–102.
Green, W.G., *Theory of Machines*, Blackie & Sons Ltd., UK, 1955.

Chapter 5
Dowson, D. and Higginson, G.R., *Elasto-Hydrodynamic Lubrication*, Pergamon Press Ltd., Oxford, 1977, ISBN 0-0802-1303-0.

Chapter 6

Steel, concrete and composite bridges. Part 10: Code of practice for fatigue. BS 5400:1980.

Code of practice for fatigue design and assessment of steel structures. BS 7608:1993.

Guide on methods for assessing the acceptability of flaws in fusion welded structures. BS 7910:1999.

Chapter 7

Plastic pipe for the conveyance of fluids under pressure. BS EN ISO 13760:2000.

The British Standards cited in Chapter 6 also refer to Miner's rule.

Chapter 8

Reliability of systems, equipment and components. BS 5760:Part 2:1994.

Wang, W. (ed.), *Process Machinery-Safety and Reliability*, Mech. Engng. Publns. Ltd., Bury St Edmunds, UK

Thompson, G., *Improving Maintainability and Reliability through Design*, Mech. Engng. Publns. Ltd., Bury St Edmunds, UK.

Davidson, J. and Hunsley, C. (eds), *The Reliability of Mechanical Systems*, 2nd edn, Mech. Engng. Publns. Ltd., Bury St Edmunds, UK, ISBN 0-8529-8881-8.

Harris, J., "Piecewise Linear Reliability Data Analysis with Fuzzy Sets", *Proc. Instn. Mech. Engrs.*, Vol. 215 Part C, 2001, pp. 1075–1082.

Chapter 9

Rowlands, H.S. and Wang, L.R., "An Approach of Fuzzy Logic Evaluation and Control in SPC", *Qual. and Reliab. Engng. Int.*, Vol. 16, 2000, pp. 91–98.

Harris, J., "On the Correlation of Statistical and Automatic Process Control", *Proc. Instsn. Mech. Engrs.*, Vol. 217, Part B, 2002, pp. 1–11.

Harris, J., "On System Condition Auditing", *Trans. I. Chem. E.*, Vol. 8, Part B, 2002, pp. 197–203.

Chapter 10

Butterworth, M., "The Emerging Role of the Risk Manager", *Mastering Risk*, Part 1, Financial Times Supplement (UK), April 25, 2000, pp. 12–13.

Glasserman, P., "The Quest for Precision through Value-at-Risk", ibid., Part 4, May 16, 2000, pp. 6–7.

Fenton-O'Creevy, M. and Soane, E., "The Subjective Perception of Risk", ibid., Part 1, April 25, 2000, pp. 14–15.

Davidson, J. (ed.), *The Reliability of Mechanical Systems*, 2nd edn, Mech. Engng. Publ., UK, 1994, ISBN 0-8529-8881-8.

Andrews, J.D. and Moss, T.R., *Reliability and Risk*, Longman Scientific and Technical, UK, 1993, ISBN 0-5820-9615-4.

Margenau, H., "Probability (Physics)", in *McGraw-Hill Encyclopedia of Science and Technology*, 19, Vol. 5, pp. 401–404.

Wong, W. (ed.), *Process Machinery-Safety and Reliability*, Mech. Engng. Publ., UK, 1997, ISBN 1-8605-8046-7.

Knight, R., and Pretty, D., "Philosophies of Risk, Shareholder Value and the CEO", ibid., Part 10, June 27, 2000, pp. 14–15.

Harris, J., "Piecewise Linear Reliability data Analysis with Fuzzy Sets", *Proc. Instn. Mech. Engrs.*, Vol. 215, Part C, 2001, pp. 1075–1082.

Harris, J., *An Introduction to Fuzzy Logic Applications*, Kluwer Academic Publishers, Dordrecht, 2000.

Carey, A. and Turnbull, N., "The Boardroom Imperative on Internal Control", ibid., Part 1, April 25, 2000, pp. 6–7.

Lewin, C., "Refining the Art of the Probable", ibid., Part 2, May 2, 2000, pp. 2–4.

Chapter 11

Geiker, M., "Durability Design of Concrete Structures", in *Minimising Total Life Costs – Considerations and Examples*, Proc. Intnl. Conf. on Concrete Durability and Repair Technology, September 1999, Univ. Dundee, Scotland, R.K. Dhio and M.J. McCarthy (eds), Th. Telford, UK, 1999, pp. 319–333.

Harris, J., "Fuzzy Logic Methods in Fatigue and Creep", *Jour. Strain Anal.*, Vol. 36, No. 4, 2001, pp. 411–420.

Harris, J., "Piecewise Linear Reliability Data Analysis with Fuzzy Sets", *Proc. Instn. Mech. Engrs.*, Vol. 215, Part C, 2001, pp. 1075–1082.

Varona, J.M., Gutierrez-Solana, F., Gonzalez, J.J. and Gorrochategui, I., "Fatigue of Old Metallic Railroad Bridges", in *Intnl. Symp. on Fat. Design*, J. Solin et al. (eds), ESIS Publication 16, Mech. Eng. Publ. Ltd., London, 1993, pp. 121–135.

Wang, L.R. and Rowlands, H., "The Evaluation and Control of SPC in Fuzzy Logic and Neural Networks", in *2nd Workshop on Euro. Sci. and Indust. Collaboration (WESIC'99)*, Newport, Gwent, UK pp. 391–398.

Harris, J., "Reframing Standards Using Fuzzy Sets for Improved Quality Control", *Proc. Instn. Mech. Engrs.*, Vol. 16, 2001, pp. 91–98.

Glasserman, P., "The Quest for Precision through Value-at-Risk", in *Mastering Risk, Financial Times*, Part 4, May 16, 2000, pp. 6–7.

Harris, J., *An Introduction to Fuzzy Logic Applications*, Kluwer Academic Publishers, Dordrecht, 2000.

Davidson, J. (ed.), *The Reliability of Mechanical Systems*, Mechanical Engineering Publications, London, 1994.

Andrews, J.D. and Moss, T.R., *Reliability and Risk Assessment*, Longman Scientific and Technical (now Pearson Education Ltd., London), 1993.

Chapter 12

Chen, Y.H., "Fuzzy Ratings in Mechanical Engineering Design – Applications to Bearing Solutions", *Proc. Inst. Mech. Engrs.*, Vol. 210 (B1), 1996, pp. 49–53.

Hinduja, S. and Xu, X., "Determination of Finishing Features in $2\frac{1}{2}$-D Components", *Proc. Inst. Mech. Engrs.*, Vol. 211 (B2), 1997, pp. 125–142.

Lin, J-J., Lin, C-H. and Tsai, I.S., "Applying Expert System and Fuzzy Logic to an Intelligent Diagnostic System for Fabric Inspection", *Textile Research Jnl.*, Vol. 65 (12), 1995, pp. 697–709.

Harris, J., "Raw Milk Grading Using Fuzzy Logic", *Intnl. Jour. of Dairy Techn.*, Vol. 51 (2), 1998, pp. 54–56.

Fletcher, M.J., "Non-Destructive Testing", in *Kempes Engrs. Year-Book*, 1999, D2/1089-1106, ISBN 0-8638-2408-0.

Harris, J., Private Communications to the British Standard Institution dated 27/11/99 and 31/11/99.

Halmshaw, R., *Non-Destructive Testing*, Edward Arnold, 1987, ISBN 0-7131-3634-0.

Ross, J.R., *Fuzzy Logic with Engineering Applications*, McGraw-Hill Inc., 1995, ISBN 0-0705-3917-0.

Harris, J., *An Introduction to Fuzzy Logic Applications*, Kluwer Academic Publishers, Dordrecht, 2000.

Chapter 13

Raadnui, S., "Condition Monitoring of Earth Moving Equipment – A Novel Approach", in *3rd Intnl. Conf. on Quality, Relaibility and Maintenance*, G.J. McNulty (ed.), Prof. Engng. Publ. Ltd., Suffolk, UK, 2000, pp. 159–162, ISBN 1-8605-8256-7.

Kreysig, E., *Advanced Engineering Mathematics*, 3rd edn, Wiley and Sons Inc., New York/London, 1972.

Harris, J., *An Introduction to Fuzzy Logic Applications*, Chapter 5, Kluwer Academic Publishers, Dordrecht, 2000, ISBN 0-7923-6325-6.

Rowlands, H. and Wang, L.R., "An Approach of Fuzzy Logic Evaluation and Control in SPC", *Qual. and Relaib. Engng. Intnl.*, Vol. 16, 2000, pp. 91–98.

Wang, L.R. and Rowlands, H., "A Fuzzy Logic Application in SPC Evaluation and Control" in *7th IEEE Intnl. Conf. on Emerging Tech. and Factory Automation*, Barcelona, Spain, pp. 679–684.

Davies, O.L., *Statistical Methods in Production and Research*, Oliver and Boyd, London, 1958.

Thompson, S., "Control Systems Engineering and Design", Longman Scientific and Technical, Harlow, UK, 1989, ISBN 0-5829-9468-3.

Reznik, L., *Fuzzy Controllers*, Newnes, Oxford, UK, 1997, ISBN 0-7506-3429-4.

Davidson, J. (ed.), *The Reliability of Mechanical Systems*, 2nd edn, Chapter 13, Mech. Engng. Publ. Ltd., London, 1994, ISBN 0-8529-8881-8.

Raz, T., "Control Chart", in *McGraw-Hill Encycl. of Science and Tech.*, Vol. 4, McGraw-Hill, London/New York, 1997, pp. 405–410, ISBN 0-0791-1504-7.

Raz, T., "Quality Control", in *McGraw-Hill Encycl. of Science and Tech.*, Vol. 14, McGraw-Hill, London/New York, 1997, pp. 645–649, ISBN 0-0791-1504-7.

Chapter 14

Harris, J. and Kanhukamwe, Q.C., "Similarity Treatments of Group Technology", in *Proc. 3rd Intn. Conf. on Manuf. Processes, Systems & Opns. Man. in LIRs*, Nat. Univ. Sci. & Tech., Bulawayo, Zimbabwe, 1998, pp. 62–67, ISBN 0-7974-1794-X.

Ross, T.J., *Fuzzy Logic with Engineering Applications*, McGraw-Hill Inc., 1995, ISBN 0-0711-3637-1.

Harris, J., *An Introduction to Fuzzy Logic Applications*, Kluwer Academic Publishers, Dordrecht, 2000, ISBN 0-7923-6325-6.

Urry, S., *Introduction to Operational Research*, Longman Scientific & Technical, 1991, ISBN 0-5820-1349-6.

Index

International Series on
MICROPROCESSOR-BASED AND INTELLIGENT SYSTEMS ENGINEERING

Editor: Professor S. G. Tzafestas, *National Technical University, Athens, Greece*

1. S.G. Tzafestas (ed.): *Microprocessors in Signal Processing, Measurement and Control.* 1983
   ISBN 90-277-1497-5
2. G. Conte and D. Del Corso (eds.): *Multi-Microprocessor Systems for Real-Time Applications.* 1985
   ISBN 90-277-2054-1
3. C.J. Georgopoulos: *Interface Fundamentals in Microprocessor-Controlled Systems.* 1985
   ISBN 90-277-2127-0
4. N.K. Sinha (ed.): *Microprocessor-Based Control Systems.* 1986   ISBN 90-277-2287-0
5. S.G. Tzafestas and J.K. Pal (eds.): *Real Time Microcomputer Control of Industrial Processes.* 1990
   ISBN 0-7923-0779-8
6. S.G. Tzafestas (ed.): *Microprocessors in Robotic and Manufacturing Systems.* 1991
   ISBN 0-7923-0780-1
7. N.K. Sinha and G.P. Rao (eds.): *Identification of Continuous-Time Systems.* Methodology and Computer Implementation. 1991
   ISBN 0-7923-1336-4
8. G.A. Perdikaris: *Computer Controlled Systems.* Theory and Applications. 1991
   ISBN 0-7923-1422-0
9. S.G. Tzafestas (ed.): *Engineering Systems with Intelligence.* Concepts, Tools and Applications. 1991
   ISBN 0-7923-1500-6
10. S.G. Tzafestas (ed.): *Robotic Systems.* Advanced Techniques and Applications. 1992
    ISBN 0-7923-1749-1
11. S.G. Tzafestas and A.N. Venetsanopoulos (eds.): *Fuzzy Reasoning in Information, Decision and Control Systems.* 1994
    ISBN 0-7923-2643-1
12. A.D. Pouliezos and G.S. Stavrakakis: *Real Time Fault Monitoring of Industrial Processes.* 1994
    ISBN 0-7923-2737-3
13. S.H. Kim: *Learning and Coordination.* Enhancing Agent Performance through Distributed Decision Making. 1994
    ISBN 0-7923-3046-3
14. S.G. Tzafestas and H.B. Verbruggen (eds.): *Artificial Intelligence in Industrial Decision Making, Control and Automation.* 1995
    ISBN 0-7923-3320-9
15. Y.-H. Song, A. Johns and R. Aggarwal: *Computational Intelligence Applications to Power Systems.* 1996
    ISBN 0-7923-4075-2
16. S.G. Tzafestas (ed.): *Methods and Applications of Intelligent Control.* 1997
    ISBN 0-7923-4624-6
17. L.I. Slutski: *Remote Manipulation Systems.* Quality Evaluation and Improvement. 1998
    ISBN 0-7932-4822-2
18. S.G. Tzafestas (ed.): *Advances in Intelligent Autonomous Systems.* 1999
    ISBN 0-7932-5580-6
19. M. Teshnehlab and K. Watanabe: *Intelligent Control Based on Flexible Neural Networks.* 1999
    ISBN 0-7923-5683-7
20. Y.-H. Song (ed.): *Modern Optimisation Techniques in Power Systems.* 1999
    ISBN 0-7923-5697-7
21. S.G. Tzafestas (ed.): *Advances in Intelligent Systems.* Concepts, Tools and Applications. 2000
    ISBN 0-7923-5966-6
22. S.G. Tzafestas (ed.): *Computational Intelligence in Systems and Control Design and Applications.* 2000
    ISBN 0-7923-5993-3
23. J. Harris: *An Introduction to Fuzzy Logic Applications.* 2000   ISBN 0-7923-6325-6